MECHANICAL ENGINEERING THEORY AND APPLICATIONS

TECHNICAL NOTES ON NEXT GENERATION AERO COMBUSTOR DESIGN-DEVELOPMENT AND RELATED COMBUSTION RESEARCH

MECHANICAL ENGINEERING THEORY AND APPLICATIONS

Additional books and e-books in this series can be found on Nova's website under the Series tab.

MECHANICAL ENGINEERING THEORY AND APPLICATIONS

TECHNICAL NOTES ON NEXT GENERATION AERO COMBUSTOR DESIGN-DEVELOPMENT AND RELATED COMBUSTION RESEARCH

JUSHAN CHIN
AND
JIN DANG

nova
science publishers
New York

Library of Congress Cataloging-in-Publication Data

ISBN: 978-1-53619-724-2

Published by Nova Science Publishers, Inc. † New York

Contents

Preface

About 20 years ago, Prof. A. H. Lefebvre wrote a letter to the present author, Jushan Chin. In that letter he wrote: "In the mean time, I want to congratulate you on the big success you have achieved in low-emissions combustion. I do not question your achievements. There is absolutely no doubt in my mind that you know more about gas turbine combustion (not just low emission emissions) than all of the xxxxx xxxxx people put together."

Lefebvre was my teacher and mentor. He assigned me "homework" appropriate to the time compiled during my early combustion research and experience in gas turbine combustor design-development. Thanks to Nova Science Publisher I am able to finish this homework assignment. This is a cooperative work with Dr. Dang with whom I work to finish this book.

There exist numerous books about aero engine combustion. Some doubtfully received permissions from the original authors. Few are about aero combustors; particularly aero combustor design. Aero combustors, by definition, must possess combustion, but combustion itself is not a combustor. For aero combustor design, Don Bahr wrote a chapter several decades ago, at that time was current technology, but now is out of date (thus that study is not listed in the references). From a technological development point of view there forthwith shall be a new generation of aero combustor designs. This present author, Jushan Chin, has vast experience in aero combustor design and development. Bryn Jones, a formal lead engineer chief combustion specialist wrote in a letter to me: "I have confidence in your ability as a designer. How can I overcome the politics to get a combustor of your design into our programme?" The author of this book has studied next generation aero combustor design for many years after retirement. This work will be the central part of this book.

In this book, both next generation aero combustor design-development and related combustion research will be reported.

Author

Chapter 1

Introduction to Aero Combustor Design

Aero combustor design and development are very much different pursuits from something like a house design. We never have heard of a house designed and built, then people moving in to evaluate living in the house, whereafter the house then would be modified to accommodate perceived defect, and after several attempts in this imaginary "living trial", finally people could move in and indefinitely live there. For a house in the real world, that never is the case. An aero combustor, after it has been designed and manufactured, it cannot be put into service immediately. It must be tested and modified (change of design), then retested and modified again. This process is called "*development*". So for aero combustor, such design is always combined with development. It is more correct to use the phrase "combustor design-development". But in this book, the present author will discuss design firstly; then the reader will find a separate chapter discussing combustor development.

Aero combustor design-development at this date basically is still built on combustion testing regimes, which development is quite different from the CFD function in compressor design.

Combustor design-development is related to a variety of scientific branches such as aerodynamics, fluid-dynamics, heat transfer, thermodynamics, chemical kinetics, physical-chemistry, et al. The entire design-development process shall include both aero thermal design and mechanical design. For mechanical design, factors to consider are stress analysis, lifetime prediction, material choice, weight analysis, cost engineering, etc. Thus design-development entails system engineering. In this book, the technical term "design" means aero thermal design, but will not include mechanical design.

1.1. Starting Points of Aero Combustor Design

Whenever the combustor designer needs to initiate an aero combustor design, there must be four information types available:

1.1.1. Fuel Type

For civil engine combustor, the fuel is Jet A or Jet A-1, while for military combustor, the fuel is JP-8, JP-8+ 100, or JP-10. Different fuels possess different properties, particularly high density aero kerosene fuel possesses quite different properties.

1.1.2. Engine Application

Whether the design is to be a military aero engine combustor or a civil aero engine combustor, a fighter aircraft engine combustor or a transportation aircraft engine combustor, a large civil engine combustor or a small civil engine combustor (such as thrust less than 26.7 kN), different applications will affect performance target. Low emission requirements are very different for military or civil combustor designs. The design will affect combustor lifespan requirements, while obviously military combustors cannot be expected to offer the same overhaul life expectancy as a civil combustor. The application will affect requirements such as the maximum altitude for ignition which is different for military combustors and civil combustors. The application determines the thrust level of the engine, then determines the combustor's air flow rate, pressure level, etc.

1.1.3. Requirements of Aero Combustor

Here the emphasis is on one point: the designer shall take all three aspects into consideration. They are: performance, operational and practical requirement. In some documents the operational aspect is described only as durability, which is not correct. This is because operational aspects should not be considered only from a durability perspective. It is not always true that performance is of the first priority. There are some cases that high altitude ignition is not sufficient to cause the combustor design to be modified (or completely redesigned). Many requirements are contradictory. For example, a low NO_x emission requirement requires at high power conditions more combustion air, which may affect idle lean blow out (LBO) and may affects low power condition efficiency. Combustor design must make certain compromise to offer good balances. In any case, the requirements of combustor are the starting point. Those requirements of aero combustor design will be discussed in next section in detail.

1.1.4. Engine Cycle Parameters

Engine cycle parameters are the most important foundation for combustor design. For civil engine and civil aero combustor, there are maximum power condition (i.e. 100% thrust level, or take off condition), 85% power condition (climb condition), approaching condition (30% power condition), idle condition (7% power condition) and maximum cruise condition.

For military engine and military aero combustor, there are maximum power condition, ground idle condition, and flight idle condition. There is no 85% condition, no 30% power condition, nor a maximum cruise condition. But there are cruise conditions at different Mach numbers and different altitudes.

As an example, XWB engine combustor cycle parameters are listed in following table.

Table 1. XWB engine combustor cycle parameters

parameters	100%	85%	30%	7%	cruise
Ma(lb/s)	223.6	197.4	100.5	43.2	76.72
P3t(psia)	699.3	597	262.3	100	217
T3(K)	917.9	875.3	717.2	559.5	768.1
Mf(lb/hr)	26464	21054	6787	2208	6993

For combustor design the most important parameters are:

a The total pressure at compressor exit P_{3t} is also same as combustor inlet total pressure. Then, combustor diffuser pressure loss will be deducted from this total pressure to obtain liner inlet total pressure P_{31t}

b Combustor inlet temperature T_3, usually assume liner inlet temperature is equal to T_3.

c Combustor total pressure loss coefficient $(P_{3t}–P_{4t})/P_{3t}$. P_{4t} is combustor exit total pressure and also is the liner exit total pressure. Due to the diffuser total pressure loss, there will be a liner total pressure loss coefficient $(P_{31t}–P_{4t})/P_{31t}$

d Combustor air flow rate Ma, when there is no air bleeding between inlet and liner, it is also liner air flow rate.

e Combustor fuel flow rate M_f, also is the liner fuel flow rate.

Engine cycle parameters are determined by performance department. If under very special cases, some cycle parameters are not reasonable, then the combustion engineer may propose necessary changes. For technology development, such cycle parameters are sufficient. For type engine combustor design there need more detailed cycle parameters. For example, cycle parameters are listed every 3% of the thrust interval because for this type of engine combustor there may also be a need to perform transient condition design and testing.

1.1.5. Geometrical Dimension Limitations

For type engine combustor, there are geometrical limitations. Combustor total length is limited. That ultimate the length spans the compressor outlet to turbine inlet. The combustor inlet is an annular section with its average diameter and annular height specified. The combustor exit is also an annular section, with its average diameter and annular height specified. Combustor outer casing diameter is limited. Combustor inner casing diameter is limited.

For type engine combustors, very often combustor weight must be limited. Combustor design must have a logical starting point, otherwise the designer will have to begin from a blank slate. There is one book concerning aircraft engine design. In that book, there is one chapter dealing with engine component design parameters and combustion systems (that book title and author's name are not listed in references). The author of the chapter never mentioned combustor design is based on cycle parameters. Never mentioned for type engine combustor design there must be limitations. Not ever mentioned what the requirements of combustor are.. How can a combustor actually be designed with that paucity of information?

1.2. Combustor Operational and Practical Requirements

As previously mentioned, combustor requirements include performance requirements, operational requirements, and practical usage requirements.

Number one of the operational requirements is reliability. Reliable ignition is a must, including ignition at ground level and under very low ambient temperature, such as 40°F below zero, and ignition at certain high altitude. After a flame out, a successful re-ignition at high altitude is a must (such as 30000 *ft* or 35000 *ft* altitude). Good idle condition LBO fuel air ratio (FAR), such as no flame out at FAR 0.006, is required. Sometimes good idle LBO and good high altitude ignition are also listed as performance requirements. There is no severe combustion instability. For military combustor, i.e. when an air-to-air missile is launched, the aero combustor must possess the ability to ingest exhaust gas without flame out. Reasonable maximum liner and dome wall temperature are also related to reliability, and in any direction with a one inch distance apart, the wall temperatures difference shall be less than 200°F. The flame shall stay ignited under a wide range of weather conditions, particularly under heavy rain condition. The flame must survive a water flow rate 5 times the fuel flow rate at approaching condition.

Practical requirements include that there shall be no difficulty for the manufacture, assembly and maintenance of the combustor design. Overhaul lifespan of a large engine civil aero combustor shall at least exceed 5000 cycles. For military aero combustor, the lifespan shall at least exceed 2000 cycles. Design for total lifespan should be as long as possible. Lifespan is combustor durability. Another practical requirement is combustor fuel nozzle overhaul deposition life, which shall not be shorter than the liner lifespan. The combustor shall not be overly-weighted. Cost shall not be over an allowable amount. But after the combustor has been put into service, the combustor may be modified to extend lifespan and reduce cost. Combustor must be easy to maintain. Lastly, the combustor hopefully has multiple fuel capability.

1.3. Performance Requirements

1.3.1. Combustion Efficiency

For a civil combustor, combustion efficiency requires fuel economy. For large intercontinental aircraft engine combustor, the maximum cruise condition efficiency of 99.5% is not good enough: it must be 99.9%. This 0.4% difference means a big difference in fuel saving. For military combustor, 1% combustion inefficiency means 1% more fuel consumption. Also for military combustor, after high altitude re-ignition, the combustion efficiency during this situation is directly related to the acceleration of the engine.

Nowadays for large civil engine combustor, even at ground idle condition, it is appropriate to require efficiency better than 99%. For military high fuel air ratio combustor at high power condition, efficiency must be better than 99.7%. For large civil combustor high power condition, efficiency must be better than 99.9%.

1.3.2. Total Pressure Loss Coefficient

If a combustor total pressure loss coefficient is 1% more, then that specific fuel consumption (SFC) is increased 0.5%.

The whole combustor total pressure loss coefficient should not be higher than 7%. Today's advanced combustors cannot have total pressure loss coefficient at 5% level.

1.3.3. Liner Exit Distribution

For civil combustor, exit temperature distribution quality is required. For military high FAR combustor, both the exit FAR distribution quality and temperature distribution quality are required.

The reason for such a change is that if local combustion FAR is higher than stoichiometric, its temperature can be the same as a local point combustion temperature with a FAR less than stoichiometric. Thus, one temperature distribution seems rather reasonable. Actually its FAR distribution is not good, which means that temperature distribution alone may offer some false information.

There are two parameters:

a Hot spot parameter, it is also called pattern factor, the definition is

$$Pattern\ factor = \frac{(T_{t4max} - T_{t4ave})}{(T_{t4ave} - T_{t3})} \tag{1}$$

T_{t4max} is the maximum temperature in the exit. For FAR distribution, it is called over rich factor, or (pattern factor) FAR, and the definition is

$$(Pattern\ factor)_{FAR} = \frac{(FAR_{max} - FAR_{ave})}{FAR_{ave}} \tag{2}$$

This parameter is for turbine inlet guide vanes. Both pattern factors shall be less than 0.2.

b Radial distribution profile factor

For a civil combustor, an exit temperature radial profile is required. At the exit, for temperature measurements, usually there are five radial span positions: 10%, 30%, 50%, 70%, and 90%. At each span position, one must measure exit temperature with traversing gear circumferentially all round (say, for every 6 degree one point of measurement, hence there would be 60 measurements). Then take the average of these measurements. In this way, there is an exit temperature radial profile.

Then there is an optimized exit temperature radial profile, which is provided by turbine department, and is based on two considerations. Stress is the highest at turbine root (zero span position). Higher temperature means higher stress. But high temperature gas has higher enthalpy also means a higher ability to do work. Combined these two considerations, the optimized radial profile is a curve, which has a maximum temperature point at 60% to 70% span position, and towards both top and bottom, the temperature is lower. Particularly at bottom, it is the lowest temperature position.

Then the measured radial temperature profile minus the optimized radial profile results in a radial curve, it may be negative, but also could be positive. Taking the maximum positive value divided by the average temperature rise result in a radial temperature profile factor, which shall not be more than 0.1. It is particularly not desirable to have the maximum positive difference at a rather lower span position.

For the FAR radial profile measured in a same way but with single point sampling probe. The optimized FAR radial distribution is obtained by translation of the optimized radial temperature profile into an optimized FAR radial profile, assuming at each span position the combustion is 100% efficiency. Then the measured FAR profile minus optimized FAR profile results in a curve derived from the maximum positive value divided by average FAR. This is FAR radial profile factor. It is also limited to less than 0.1.

1.3.4. Ground Idle Condition Lean Blow Out (LBO)

Ground idle condition lean blow out is a parameter for combustor flame stabilization. Usually required LBO FAR shall not be higher than 0.006.

1.3.5. High Altitude Ignition

High altitude ignition is required to have maximum ignitable altitude. For military fighter combustor, such as 35000 *ft*. But for different engines, such as helicopter or transportation aircraft, this requirement differs. For medium and large civil aero combustor the ignition altitude requirement usually is 30000 *ft*. For any rig, to run tests of simulating high altitude ignition, the simulated ignition altitude should be at least at 4 *psia* level.

1.3.6. Smoke Level

Non-visible smoke is required for all aero combustors. For civil combustor it is defined by ICAO CEAP standard (smoke requirement has not been changed for different CEAP standards). Notice that the smoke level is for all thrust level engines (even lower than 26.7 *kN*), The measured smoke number (*SN*) shall not be over 50. Or in engine certification testing, with correction for the number of testing engines, the measured *SN* shall not be over the number calculated by the following equation:

$$SN = 83.6 * (F_{00})^{(-0.274)} \tag{3}$$

where:

SN smoke number

F_{00} maximum take-off thrust

1.3.7. Gaseous Exhaust Emissions

Gaseous exhaust emissions include NO_x, CO, unburnt hydrocarbons (UHC). For engines put into service on and after 2004 the gaseous emissions are specified by CEAP 4. For engines put into service on and after 2008, the gaseous emissions are specified by CEAP 6. And from now onward, gaseous emissions are specified by CEAP 8. Notice for gaseous emissions, an aero engine company will report how much emissions reduction compared to CEAP standard. For example, say a 40% NO_x reduction relative to CEAP 8 means that in a specified take-off and landing cycle, the engine combustor will produce NO_x which is 60% of the specified value in CEAP 8. For NO_x, high power conditions are critical, but idle condition NO_x cannot be neglected either. During low power conditions, achieving low levels of CO and UHC is critical, but at high power conditions, efficiency is always high, thus CO and UHC will not be problematic.

1.4. Combustor Working Parameters

Combustor working parameters only describe the combustor's work; they do not show whether performance is good or poor.

1.4.1. Pressure

P_{3t} combustor inlet averaged total pressure, represents engine pressure ratio

P_{31t} liner inlet averaged total pressure

P_{4t} combustor or liner exit averaged total pressure

P_{3s} combustor inlet static pressure

P_{31s} liner inlet static pressure

P_{4s} combustor or liner exit static pressure

1.4.2. Temperature

T_3 combustor inlet air temperature

T_{31} liner inlet air temperature, usually assuming it is equal to T_3, also assuming no difference between total temperature and static temperature

1.4.3. Air Flow Rate

Ma_3 combustor inlet air flow rate

Ma_{31} liner inlet air flow rate, if there is no air bleeding between the inlet and the liner, then it is equal to Ma_3

1.4.4. Fuel Flow Rate

M_f fuel flow rate for the combustor or for the liner is the same

1.4.5. Fuel Injector Pressure Drop

Fuel injector pressure drop is fuel nozzle upstream pressure minus nozzle downstream (liner inside) pressure. During combustion testing, with measured fuel flow rate and measured fuel nozzle pressure drops, the working fuel injector flow number (FN) can be determined. If the working fuel injector flow number is significantly different from the new injector flow number, then the working fuel injector or the measurement must be problematic.

Fuel injector flow number is defined by:

$$FN = \frac{M_f\ (lb/sec)}{[\text{injector pressure drop } (lb/in^2)]^{0.5}}$$

1.4.6. Combustor Fuel-Air Ratio (FAR)

$\frac{M_f}{Ma_3}$

1.4.7. Liner Fuel-Air Ratio

$\frac{M_f}{Ma_{31}}$

Fuel air equivalence ratio is the working fuel air ratio over stoichiometric fuel air ratio, which is 0.068 for most commonly used aviation fuel. But for some special fuels, such as JP-10 or high density fuels, the stoichiometric fuel air ratio needs to be newly recalculated based on fuel chemical composition. Stoichiometric fuel air ratio is calculated with the assumption that in fuel composition all carbon atoms are oxidized to CO_2, and all hydrogen atoms are oxidized to H_2O, plus all sulfur atoms are oxidized to SO_2.

1.4.8. Liner Effective Flow Area AC_d

Liner effective flow area is the product of liner air flow geometrical area and discharge coefficient. It is a very important working parameter. During combustion testing if AC_d has changed significantly the liner or the test rig will experience problems.

Taking the whole liner as one air flow device, we derive the following equation:

$$Ma_{31} = AC_d * [2 * (P_{31} - P_{4s}) * \rho_a]^{0.5} \tag{4}$$

Where:

Ma_{31} liner air flow rate

AC_d liner effective flow area

ρ_a air density, defined at liner inlet condition

1.4.9. Combustion Instability Pressure Fluctuation

It is important to monitor combustion instability pressure fluctuation during combustion testing and is particularly important for lean pre-vaporized, premixed low emission combustors.

1.5. Technology Development Program Aero Combustor and Type Engine Aero Combustor Design

For these two types of combustor designs, there are some common points such as both originating from a predetermined set of cycle parameters. Both have similar requirements. For a technology development program, the target is to have new technology developed, which is suitable for integration into an engine. For a technology development program combustor, in order to develop and evaluate new technology, one also needs to run combustor tests very specific to the engine type. Certainly, there is no paradigm for a long life testing protocol. The differences between technology development program combustor design and an type engine combustor design are:

a For type engine the final outcome must be a combustor hardware, which can be installed on engine as a product that can be sold. But for technology program the final result only is to develop certain advanced technology, not necessarily a product. That is a big difference. Because of this, usually there is no mechanical design for technology development program combustor.

b Technology development programs are not for any one specific engine type. Thus there is no need to consider in detail combustor assembly. For type engine, the combustor must be suitable for one specific aero engine and it must have zero problem to assemble into an engine. But even for a technology development program combustor, if the configuration can not be assembled into an engine, it is no good.

c One big difference is that for development programs, there are always some risks. That means a chance of failure is possible. For type engine, there cannot be any failure. It is a river of no return. No matter how difficult it is, the combustor must be re-designed and developed to be successful. Without a combustor, there will be no completed engine product.

1.6. Combustor Aero Thermal Design and Mechanical Design

Aero combustor design usually is accomplished in two stages: aero thermal design and mechanical design. Always aero thermal design is the first stage, and after aero thermal design has reached a certain stage, mechanical design may begin. Mechanical design will also have final drawing for manufacturing. For next generation civil aero combustor, CMC liner, aero-thermal design, and mechanical design are closely related. At the second design stage, designers shall consider combustor material, tolerance, assembly, manufacturing method, sealing, alignment, thermal expansion analysis, vibration, stress analysis, life prediction (based on combustor test liner and dome temperature measurement and cooling calculation), etc., in details. There are also fuel manifold design and all connection design (including bolts and nuts). Fuel control system and fuel pump designs are not combustor designer's job. Even fuel nozzle detailed design and manufacturing are accomplished by fuel nozzle company. Combustor designers only determine fuel injector requirements, such as flow number, spray angle, etc. This book will only discuss aero thermal design.

Very often when people talk about combustor design, they mean aero thermal design. Usually mechanical design is not included.

Combustor aero thermal design consists of several aspects:

a Combustion organization design. This is the basic and most essential portion of the design process. For combustion organization design there are three contingencies to consider. Firstly, it is the air, air flow distribution, how much is the combustion air and how much is cooling air (for some combustors there may be a need for dilution air). Then to be considered is combustion zone aerodynamics or air flow pattern. Secondly, it is the fuel, choice of fuel injector type, determination of fuel injector flow number and spray angles, the relationship between main fuel and pilot fuel, at what operational condition main fuel is to open and how to manage pilot fuel-main fuel split after main fuel is opened, until up to the maximum power condition. Thirdly, the relationship between fuel and air. The designer shall take fuel and air as a combination to consider. Fuel spray penetration and dispersion in air flow, fuel air mixing in different operational conditions. Under certain conditions (usually high power conditions) FAR shall be uniform, while under other conditions (usually low power conditions), FAR distribution in whole liner shall not be totally uniform.

b Liner cooling design and whole combustor cooling calculations, including determination of total cooling air, division of the cooling air between outer liner, inner liner and the dome, cooling air flow distribution along axial distance, design cooling air hole configuration (this is most essential in cooling design), cooling hole axial arrangement (axially uniform or non-uniform spacing), whether incorporate thermal barrier material, combustor casing and liner cooling calculations.

c Inlet diffuser design. For advanced combustors, usually the inlet Mach number is higher than previous conventional combustor diffusers. The total pressure loss in diffuser is proportional to inlet Mach number squared. Also for next generation aero combustors, a very high proportion of combustor air is flowing directly through the dome as combustion air. Thus the diffuser requires new design considerations.

d For next generation aero combustor there is no dilution air, but for exit distribution, some design approaches must be integrated with the dome combustion design to assure suitable exit distribution.

e For next generation aero combustors, there will be no flow from the diffuser exit to the dome. But how bled air flowing to channels requires further design consideration.

For aero thermal design, there are design conditions and off-design conditions. In conventional combustion design, while always using 100% power condition as design condition, and other power conditions as off-design conditions. For the next generation aero combustors, the situation shall be changed. Pilot fuel air combustion will be designed at idle condition, while main fuel air combustion will be designed at maximum condition. Cooling may be designed at maximum condition or a 120°F hot day condition, or at a low altitude dash condition.

In this book, we will discuss brand new aero thermal design for the next generation aero combustors, including new combustion organization designs, new cooling designs, new diffuser designs.

1.7. Preliminary Design and Detail Design

Aero thermal design has two stages, preliminary design and detail design. In conventional combustor design the content of preliminary design is from starting point to the completion of a full annular liner design drawing. Basically the following work shall be done:

a Determination of total combustor air distribution

b Calculation of liner total effective flow area AC_d

c Assuming the air flow distribution is exactly the AC_d distribution, there are AC_d for different liner components

d Determination of liner cross sectional area and liner length, so there can be a liner sketch

e Determination of the channel height for outer channel and inner channel, so there can be a sketch for combustor casing f Determination of fuel injector type and the total number of fuel injectors

g Choose a cooling configuration

The preliminary design, just as A. H. Lefebvre said, was to create a general contour. It is really "preliminary".

For next generation aero combustor, the content of preliminary design shall be changed, which will be discussed in the following sections.

Chapter 2

Briefing on Next Generation Aero Combustor Design

Before the discussion of next generation aero combustor design commences, a first logical question is

2.1. What Are the Next Generation Aero Combustors?

Aero gas turbine engine has seen development for over 80 years. Whether it is a civil engine, or a military engine, they are all developed under one general goal towards higher performance, higher reliability and lower fuel consumption. In 1977, the International Civil Aviation Organization (ICAO) published a document entitled "Control of Aircraft Engine Emissions". Since then aero combustor entered a new era of low emissions combustor, but the requirements of low emissions are very different for civil aero combustors and military aero combustors. For civil aero combustor, the emissions are regulated by ICAO Committee on Aviation Environmental Protection (CEAP). The standard was developed from CEAP 1, CEAP 2, CEAP 4, CEAP 6, and CEAP 8. This new standard is becoming more and more strict. Nowadays any civil aero engine of thrust higher than 26.7 *kN* must be in compliance with the standard. Their emission of NO_x, CO, unburnt hydrocarbon (UHC), and smoke shall be controlled.

Because of the requirement of continuous improvement for reduction of fuel consumption, civil engine are developed with two aspects considered. An aero engine as a propulsion unit, the design effort is driven by propulsion efficiency.

Improvements of propulsion efficiency are accomplished by increasing engine's bypass ratio. From another perspective, an aero engine is also a thermal machine requiring high thermal efficiency. The way to improve thermal efficiency is to increase pressure ratio (and at the same time increase turbine inlet temperature accordingly). For several decades, aero engine pressure ratio has been increasing constantly. From nearly 10, then upward to 20, 30, 40, and 50, while a pressure ratio 60 civil aero engine (GE9X) has been put into service. The next generation civil engines will feature extra high pressure ratio of 70 and above.

Thus, the next generation civil aero combustor will be a high pressure, low emissions combustor.

For military aero engine, the major development is higher thrust-to-weight ratios. To improve thrust-weight ratios, the engine must feature higher turbine inlet temperatures (or higher FAR) and increase the engine pressure ratio accordingly. The combustor FAR has been increased from lower than 0.02 to 0.03, to 0.038, to 0.046. Next generation military engine will have extra high FAR (0.051 and above).

Next generation aero combustors are high pressure, low emissions civil aero combustors, and high FAR military aero combustors.

2.2. High Pressure Civil Aero Combustor Can Not Use LPP Combustion

LPP (lean pre-vaporized premixed) combustion must feature a premixing fuel/air module wherein the fuel and air premixed; then the mix enters the liner combustion zone. The time fuel and air remains in the pre-mixer is residence time. Fuel will vaporize in the heated air, then fuel vapor and air form the flammable mixture. If the heat generated from the exothermic pre-ignition chemical reaction can not dispersed outwardly (as is always the case), then heat will accumulate in the system and the system temperature will rise. With increased temperature the pre-ignition chemical reaction accelerates. Finally the system will experience auto-ignition. From fuel injection into the pre-mixer up to the occurrence of flame, this is defined as the auto-ignition delay time. Obviously, the fuel-air residence time in the premixing module must be shorter than the auto-ignition delay time with an appropriate safety factor. A. H. Lefebvre proposed a safety factor of 5, which in the formative years of low emissions combustor development, design was not based on good experimentally-derived auto-ignition delay time data. For more a modern combustor design, the present authors proposed a safety factor of 2. Auto-ignition time for aviation fuel at high pressure and high air temperature was measured by experiments (Guin 1999). This is exceptional experimental research by ONERA showcased auto-ignition delay time measured in conditions up to 30 *atm* and 900°K. Then there also is a correlation equation to predict auto-ignition delay time. The correlation is as follows:

$$t_{auto} = 0.508 * \exp(3377/T) * P^{-0.9} \tag{5}$$

where

t_{auto} auto-ignition delay time, ms

T air temperature, K

P air pressure, bar

As shown by equation 5, auto-ignition delay time is almost inversely proportional to pressure and exponentially decreasing with an increase of temperature (mainly chemical delay time is exponentially decreasing with temperature). At high pressure and high air temperature the auto-ignition delay time will be decreased to very low values. For example, by equation 5, we may have, under a pressure of 42.5 *bar* with an air temperature of 900°K, the auto-ignition delay time is 0.74 *ms* with a safety factor of 2, and the usable pre-mixer residence time is 0.37 *ms*. But such correlation equation (experimental data were obtained under pressure of 30 *bar*) can not be extrapolated to predict a 70 *bar* combustor inlet case because it is too far from the experimentally tested conditions. Also it is impossible to conduct auto-ignition experimental testing to determine auto-ignition delay time under 70 *bar* combustor inlet condition. The present author did a research, then proposed a calculation method. The method include physical delay time, which is from the initiation of fuel injection to the formation of a flammable mixture. And the chemical delay time, which was from flammable mixture formation to occurrence of a flame. Assuming the pre-ignition chemical reaction is a simple one step reaction, the heat release is assumed as the fuel heating value, with the chemical reaction rate constant derived from the data in Guin (1999). Then we used this calculation method to predict an auto-ignition delay time under 70 *bar* combustor inlet situation, which delay time is 0.31 *ms*, with a safety factor of 2, and a usable premixing module residence time is 0.155 *ms*. It is obvious that for a 70 *bar* pressure combustor inlet condition there is no way to use pre-vaporized premixed combustion. Not only it is inherently too dangerous, also there really is no advantage to induce such a short timeframe for premixing.

Lean pre-vaporized premixed combustion cannot be used, then what is the way out?

2.3. LDM Combustion Not LDI Combustion

More than twenty years ago, some one suggested using lean direct injection (LDI) combustion for a low emissions combustor. This suggestion was simple, and only a sketch. Later on there emerged an extended definition of LDI, but such definition still did not propose specific design approaches, how to achieve low emissions (not listed in the references). Actually direct fuel injection was not a new concept even at that time. All conventional aero combustors (except vaporizer) had fuel directly injected. There was no explanation how and what the designer should do to achieve low emissions: no strategy nor any appropriate design configurations. There were several LDI fuel air module configurations tested. Unfortunately the expected lowered emissions were not achieved. Even combustion efficiency was not improved. Even more troublesome is the fact that these fuel/air module configurations cannot be integrated onto an engine combustor. For the newest combustor technology, there must be a concerted development strategy. For example, for low emissions combustor, how to solve idle LBO problem, how to solve the high power condition EI NO_x problem, how to solve the high altitude ignition problem, and how to assure transient operation from low power condition to high power condition with fuel staging or without fuel staging, etc. Only then by combining these strategies into one overall design will ultimate efficiences be achieved. Without such strategies and overall considerations, the research on individual issues, such as swirling strength, fuel atomization, etc., will not generate the final optimal combustor design.

In recent years the present author proposed a concept, called lean direct mixing combustion (LDM). This concept is applicable for high pressure civil aero combustors. It cannot be used for premixing combustion. It offers the potential for lowered emissions through enhanced fuel/air mixing. This is not premixing, rather it is direct mixing combustion. For any high power low emissions combustor, direct mixing (LDM) offers the potential for lowered emissions and enhanced fuel efficiences.

This concept has emphasized that the major design approach shall be concentrated on how to improve high power condition fuel/air direct mixing. The present authors propose two detailed designs for direct mixing fuel air modules, for which the emissions are lowered. In one design there is no fuel staging. For the alternative design there is fuel staging. The present authors will introduce such designs later. For each design there is an overall strategy. At this moment there is only a conclusion:

For high pressure civil aero combustors, lean direct mixing (LDM) combustion and not lean direct injection (LDI) combustion is the optimal development approach.

2.4. High FAR Combustor Needs Direct Mixing Combustion but Not Lean

For any high FAR combustor premixing combustion can never be used. What still is needed is enhanced fuel/air mixing. But not lean mixtures. At maximum power conditions, it is stoichiometric direct mixing combustion that should offer optimal efficiences. It is interesting to note that the direct mixing fuel air module for the first type high pressure civil combustor (without fuel staging), and for high FAR combustor, are basically of the same configuration – of course, with different design choices and different geometrical dimensions.

2.5. Next Generation Aero Combustors Need Very High Combustion Air Fractions

For high FAR combustor, say, of a FAR 0.051 at maximum power condition, the combustion equivalence ratio shall be unity and combustion shall be uniform. For commonly used aviation fuel the stoichiometric fuel/air ratio is 0.068. For combustor FAR 0.051, using 75% combustion air, the combustion FAR will be $0.051/0.75 = 0.068$. This is stoichiometric as design required.

For a civil aero combustor, suppose its maximum condition FAR is 0.034, using 80% combustion air, at maximum condition the combustion FAR is $0.034/0.8 = 0.0425$. Its equivalence ratio is 0.625. We know for low NO_x emissions the combustion equivalence ratio shall be lower than 0.7. Because of very high pressure without premixing, the combustor must achive leaner combustion, and under such conditions the combustion also must be reliable.

Because of a very high combustion air fraction, the whole liner layout must be changed.

2.6. Next Generation Aero Combustor Layout

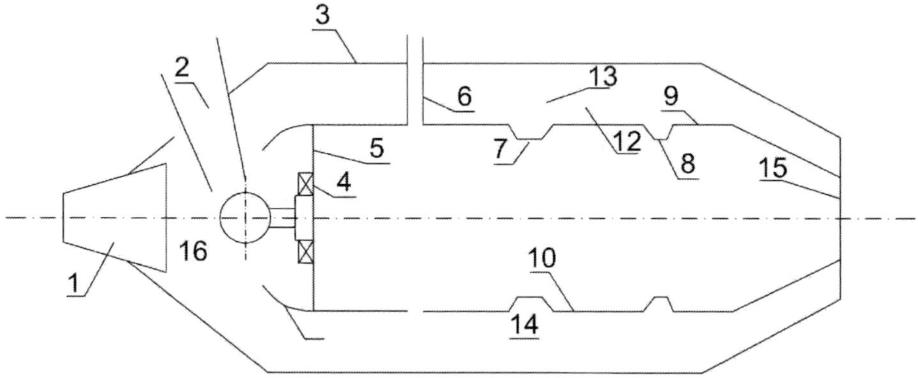

Figure 1. Conventional combustor layout: 1. Diffuser 2. Fuel nozzle 3. Casing 4. Dome swirler 5. Liner dome 6. Ignitor 7. Primary air hole 8. Dilution hole 9. Outer annular liner 10. Inner annular liner 11. Dome snout 12. Cooling hole 13. Outer channel 14. Inner channel 15. Combustor exit 16. Dumping zone.

The layout sketch of conventional aero combustor is shown in Fig 1. Note that there are primary air holes, intermediate air holes and dilution air holes. Also note that there is a dump diffuser.

Fig. 2 shows the layout sketch of next generation aero combustor. There are no liner entry air holes. In the front there is an air bleeding diffuser directly in contact with the liner dome. The bled air flows to the outer and inner channels for cooling air.

Why there shall not be any primary air holes?

According to conventional combustor design practice, half of the air passing through primary holes (supposed) will contribute to combustion air (actually combustion air is less than 50%, probably about 40%), while the remained flows downstream. If the primary holes are still used, suppose there needs to be 75% of combustion air, then a 50% combustor air shall pass through dome and 50% combustor air shall pass through primary holes (dome 50% plus 0.5 $*$ primary

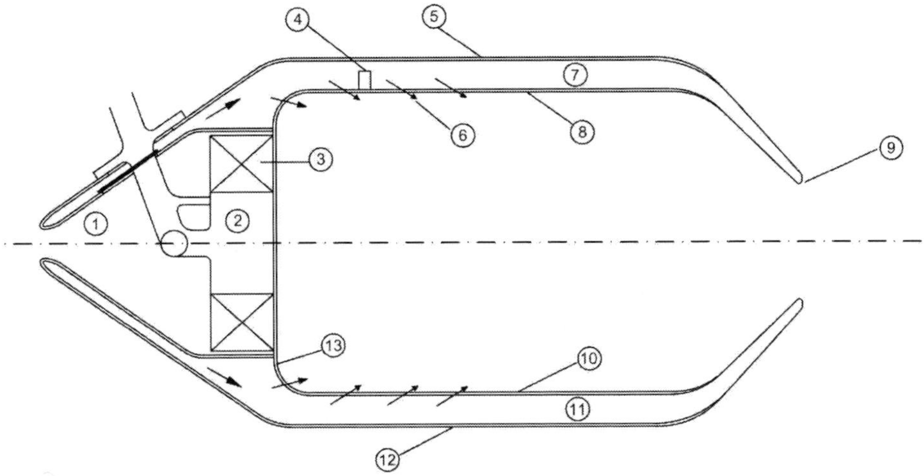

Figure 2. New generation combustor: 1. Diffuser 2. Fuel injector and pilot air module 3. Main air module 4. Ignitor 5. Outer air casing 6. Cooling air 7. Outer channel 8. Outer annular liner 9. Combustor exit 10. Inner annular liner 11. Inner channel 12. Inner air casing 13. Liner dome.

hole 50%, together makes 75%). So there is no cooling air. This is impossible for a combustor. That means, with only one way out, 75% of combustion air all comes in through the dome. Because of such air distribution, there is no primary hole air, nor any dilution air. That is a critical change. Also, using 75% and more combustion air means cooling air can only be 25% or less. There needs to be a new cooling technology (This is particularly true for high FAR combustor). As reported in Bahr (1987), a high combustion air fraction will cause idle LBO problem. Considering all these modifications to aero combustor design together, the next generation of aero combustors will feature a very different technology derived through very different design methods.

Why have the present authors identified that there should be a new generation of aero combustor designs?

For aero combustors, except during the "infant" development period (there was no definite overall design configuration), and so far there only have been two generations of aero combustor design:

a First generation features are: a faired (or aerodynamic) diffuser, dual orifice fuel injectors, three axial zones (primary zone, intermediate zone, and dilution zone), and within the liner there are three groups of air entry hole, liner cooling is from wigglestrips or machined ring, the combustor is relatively long, while the primary combustion zone is of one overall combustion zone. Many combustors are can-annular combustors.

b Second generation features are: a short dump diffuser, an air blast atomizer, no intermediate entry holes, machined ring cooling or effusion cooling, and while combustor length is shortened it is a short annular combustor. Many combustors are annular combustors.

c In the near future, there will be a new generation of aero combustors. The salient features of such are: an air bleeding diffuser, direct mixing combustion, two concentric combustion zones (pilot combustion zone and main combustion zone), pressure swirl nozzles for pilot fuel combined with axial air swirler for pilot combustion, co-flowing air with angled plain jet tube injector for main fuel combustion, very high combustion air fraction, and on the liner there are only cooling air holes, and a compound angle tangential inlet cooling hole configuration. Mostly these will be annular combustors.

These salient features will be discussed in the following sections.

Chapter 3

Type One High Pressure Low Emissions Civil Aero Combustor (without Fuel Staging) Preliminary Design

For next generation aero combustor (not just low emissions combustor), the preliminary design should focus on a single fuel air module tubular combustor, and to test and develop this tubular combustor and so solve the basic combustion organization problem. This is a big change from the conventional combustor preliminary designs. To start with, there are two things firstly to determine: combustion air fraction (say, 75%) and liner combustion air AC_d, plus the choice of total fuel-air module numbers to determine single module tubular combustor diameters.

3.1. Determination of Single Module Combustor Liner Diameter

Taking a known liner total pressure loss coefficient plus 0.3% (based on experience) to have liner front side total pressure minus exit static pressure P_{31t}–P_{4s}. Assuming at liner dome side, the static pressure is equal to total pressure, using T_3 to calculate air density, plus factoring the required combustion air fraction (then combustion air flow rate) and using Equ. 4 to calculate combustion air AC_d. One can calculate the full annular liner cross section area by 12 times the combustion air AC_d.

Choose a number of fuel air modules. Designers will prefer to use numbers such as 12, 16, 20, etc., while numbers such as 15 and 18 are not preferred, and number 13 and 17 are never used. The full annular cross sectional area divided by the number of total fuel air modules is the tubular liner cross sectional area, to then allow determine of its diameter. According to present authors' experience, the tubular liner may be obtained from a standard Hastelloy X pipe by machining its outer surface and inner surface to reach the desirable diameter and wall thickness 0.08 *in*. Tubular liner length is 8 *in*. For any CMC full annular liner, a single module tubular liner is also a CMC tubular liner.

3.2. Fuel Air Module Configuration

For a low emissions combustor without fuel staging, a fuel/air module configuration is shown in Fig 3, Chin and Dang (2021). The pilot fuel/air module is in the center, the main air module surrounds the pilot fuel/air module, and they are concentric. Pilot fuel nozzle is in the center of pilot module, and the main fuel injector is incorporated with a pilot fuel/air module. In the engine combustor the pilot fuel nozzle, main fuel injector and pilot air module comprise just one piece, installed from the air casing. Main fuel is injected at an angle relative to the module center line. This angle is a critical design feature in this combustor. the pilot module is a curved blade axial air swirler with a convergent exit. The main air module inlet comprise two portions, one for non-swirling air and the other for swirling air. The swirling air passes through a curved axial swirler. Non-swirling air intakes the inner part of main air module. Between the pilot air module exit diameter and the main air module exit inner diameter there is a separation distance.

Pilot air module has little cooling air and main air module also has little cooling air on its inner side.

3.3. Idle Condition Design

The design goal is to determine how much air is used for pilot fuel combustion at idle condition. During an idle condition the main fuel is not enabled, only

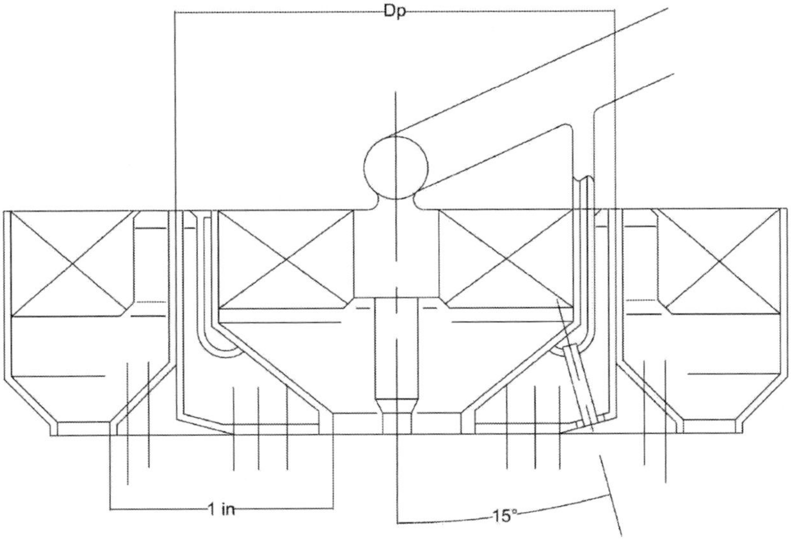

Figure 3. Fuel air module for extra high pressure low emissions combustor without fuel staging.

pilot fuel is operational. The idle condition pilot nozzle fuel flow rate is equal to idle fuel flow rate.

This combustion design procedure is very different from previous conventional design procedure. Now the procedure starts from the idle condition design and the pilot fuel air combustion is designed at idle condition.

For single module pilot fuel nozzle, its fuel flow rate is the idle condition fuel flow rate divided by the total number of fuel air modules. Design idle condition pilot fuel air combustion at equivalence ratio 1.3 to determine idle condition pilot combustion air flow rate, or pilot air fraction of total combustion air (usually assuming that the air division between pilot and main is the same at all operational conditions). This design choice is to optimize idle condition NO_x, idle condition LBO and idle condition combustion efficiency. Notice that the pilot module air is not equal to the pilot combustion air: the difference is a small amount of cooling air.

3.4. Design Air Flow Pattern

The designer may confront five possible considerations:

- pilot air swirlers swirling angle between 35° to 45°
- in the main air module, the ratio of non-swirling air over main module air should range between 25% to 33%
- main air swirling angle of between 45° to 60°
- pilot combustion air and main combustion air division

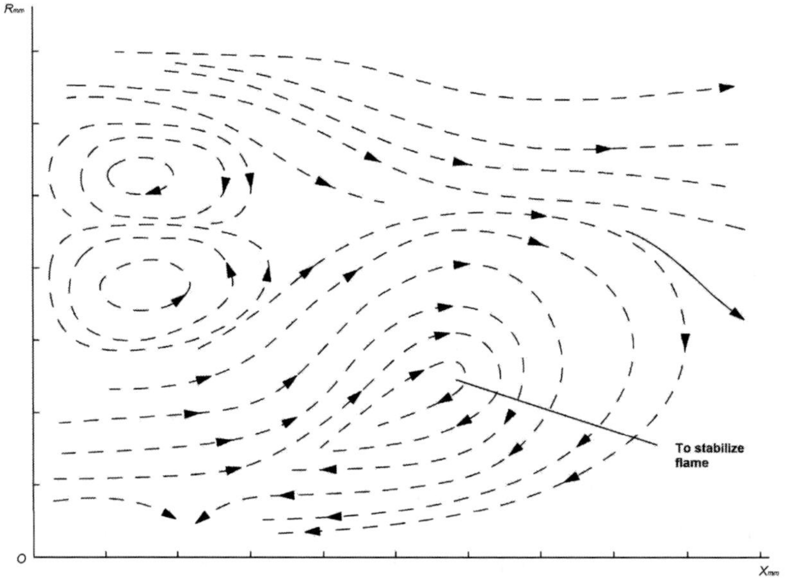

Figure 4. Air flow pattern.

The purpose is to have an air flow pattern which shows that during idle condition the main air flow is just passing by the pilot air recirculation zone, as shown in Fig 4. The present authors' experience is to start from a baseline, such as:

Pilot swirler 45°.

Main module air non-swirling 25%

Separation distance 1.0 *in*

Pilot combustion equivalence ratio 1.3

That means both pilot module and main module shall have been aerodynamically designed.

By CFD or by laser diagnostic to obtain an air flow pattern similar to that in Fig. 4 with the local small recirculation removed. Then using the same CFD to predict idle condition air flow pattern (usually not much different from the atmospheric flow pattern). When there is combustion, the air flow pattern (hot pattern) will not be the same as the cold pattern. But there is no need to verify the hot air flow pattern. Because the designer can directly run an idle LBO test. It is very simple to judge whether the idle condition air flow pattern is good for idle LBO. Note that within the above mentioned five factors, the pilot fuel/air combustion equivalence ratio (which means pilot combustion air flow division), cannot easily be changed, and that will directly and significantly affect idle LBO. Another two design choices will effectively affect the flow pattern: one is pilot air swirling angle, and the other one is the separation distance. There is no constantly correct answer. The designer must choose a baseline then start the design optimization.

Pilot module design also will affect idle LBO. Pilot module design starts from the determination of pilot module AC_d. Notice that pilot module air flow is pilot fuel combustion air AC_d minus pilot module cooling air AC_d (although only a small amount). Then based on previously determined liner exit static pressure (assuming it is equal to dome side static pressure), plus the liner inlet total pressure, air density and pilot module air flow rate, we are able to calculate the pilot module AC_d. This is pilot module required AC_d.

Pilot module inlet is a curved, axial flow swirler with a blade thickness of 0.04 *in*. The inlet AC_d is designed at 1.8$*$pilot module required AC_d. Pilot exit AC_d is calculated by the empirical equation as follows:

$$AC_{d\ pilot\ req} = AC_{d2} * [1 + (AC_{d2}/AC_{d1})^2]^{-0.5} \tag{6}$$

where

AC_{d1} is the pilot module inlet swirler AC_d

AC_{d2} is the pilot module exit AC_d

This empirical equation is based on simplification that the air exiting the swirler has lost dynamic head, and at the same time, neglecting the swirling in exit air flow. This simplified equation has been proven useful for relatively weak swirling flow.

The pilot module exit area shall be used to consider the fuel nozzle exit area, with an exit geometrical area of $1.02 * AC_{d2}$ to determine pilot module exit diameter.

Idle condition pilot fuel nozzle pressure drop has significant effect on idle LBO. The pilot fuel nozzle is a simple pressure swirl nozzle. It is well known that under low nozzle pressure drop a pressure swirl nozzle offers better atomization than any other type of fuel injector. During idle condition, by reducing FAR the nozzle pressure drop is decreased proportionally to the decrease of fuel flow rate squared. It is desirable that at idle flame out the pressure swirl nozzle still has more than a 30 *psig* pressure drop to keep the nozzle spray above an onion atomization mode (trumpet mode). A pressure swirl nozzle features pencil mode when the pressure drop is lower than 16 *psig*. When pressure drop is between 16 to 28 *psig*, it is onion mode. With a pressure drop above 30 *psig* but lower than 50 *psig*, it is trumpet mode. With plenty of flowing air, at a nearly flame out moment, pilot fuel atomization is in an air blast atomization mode which is beneficial to delay flame out. Because of this, at idle condition, design pilot fuel nozzle pressure drop at 180 *psig* to determine pilot fuel nozzle flow number (FN). A spray angle of 90° is a compromise between idle LBO and high altitude ignition. Note that the pilot module air swirler swirling angle should be a little bit less than half of the spray angle.

To summarize, the pilot fuel air combustion design is concentrated on idle LBO, while there are also considerations of other aspects. For instance, for this design, there are several major factors:

- pilot fuel air combustion equivalence 1.3
- air flow pattern design
- pilot air swirling is relatively weak swirling, with a swirling angle only a little bit less than the half spray angle
- idle pilot fuel nozzle pressure drop designed at 180 *psig*
- pilot module exit diameter and main module exit inner diameter separation distance is one inch

Although the pilot fuel air combustion design is concentrated on idle LBO, all these design considerations are beneficial for idle efficiency and idle NO_x. Experiences have proven these design considerations are correct.

Pilot fuel/air combustion design uses a combination of simple pressure swirl nozzle and an axial air swirler. This is a very good combination. At low power conditions this design is beneficial for idle LBO, idle efficiency, idle EI NO_x, and 30% power condition flame stabilization. At high power condition, this design's EI NO_x is rather close to a premixed combustion design. Because of this situation, the present authors proposed an alternative design for a high pressure, low emissions combustor featuring fuel staging, and using this combination for the entire combustor. That will be reported later in this book.

3.5. Main Fuel Air Combustion Design

Main fuel air combustion is designed for maximum power condition. There is an assumption that the pilot air and main air division at a maximum power condition is the same as at an idle condition (this is only approximately correct). Because, for the following three situations, this air flow division is not totally the same:

- under atmospheric conditions, the air division is evaluated by an air distribution test
- at idle condition pilot fuel is burning, but no main fuel is burning
- at maximum condition, both pilot fuel and main fuel are burning

For these three situations the air flow split is not totally the same. But for combustor design purpose this assumption may be used.

The present authors introduce two combustor engineering concepts: AC_d is not a constant. Fuel nozzle flow number is not a constant. These two concepts are important for combustor design-development. The liner AC_d is changing with FAR, More correctly it is a function of FAR/T_{31}. Under different conditions liner AC_d variation is shown in Fig. 5. As shown in Fig. 5, the number one condition is under high pressure high FAR combustion with heated air (say, maximum power condition). Number two is under high pressure heated air but no combustion. Number three is under ambient pressure non-heated air, no combustion; very often this is called cold AC_d. AC_{d1} is less than AC_{d2}, and clearly

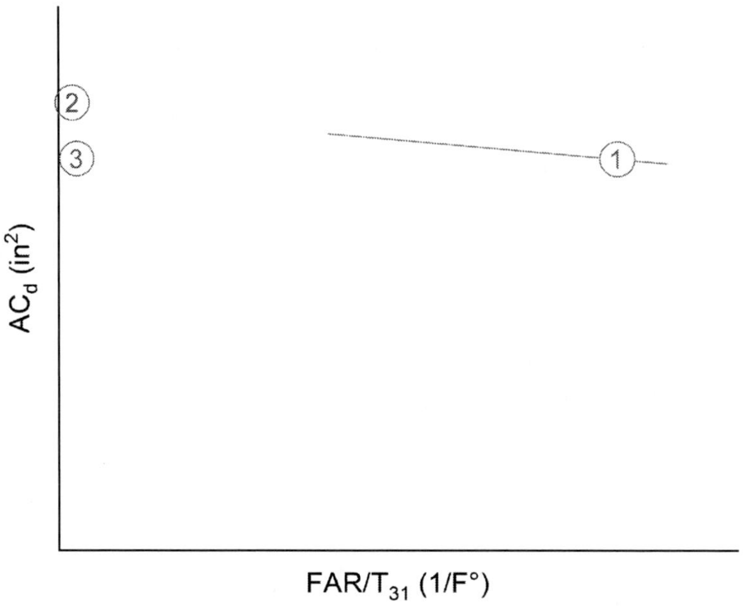

Figure 5. AC_d changes with (FAR/T_{31}) (1) under high pressure high FAR combustion with heated air, (2) under high pressure heated air but no combustion, (3) ambient pressure non-heated air no combustion.

shows that AC_d is decreasing with increasing of FAR. AC_{d3} is less than AC_{d2}. An interesting point is that very often AC_{d1} is rather close to AC_{d3}. Under atmospheric condition the designer has measured the liner AC_d which is rather close to the maximum power condition operational AC_d. The dotted line represents the changing slope of AC_d variation.

AC_d of air flow components are all non-constant. They are functions of *Re* number. Because C_d is a function of *Re* number, on the liner wall and dome there air flowing components under high metal temperature, wherein even geometrical area is not a constant.

With all these explanations, it is still reasonable, for engineering purpose, to assume that the pilot air AC_d and main air AC_d split under idle condition and under maximum condition, are the same.

Fuel nozzle flow number is also not a constant, because it changes with fuel flow rate or nozzle pressure drop (*Re* number), and decreasing with nozzle interior deposition or nozzle exit deposition.

Such understanding is very useful when running combustor testing to judging whether the combustor is working properly or some unusual situation has happened.

This paragraph is to illustrate that for combustor design, some sort of engineering approximation is necessary. It depends on the designer's experience.

Main fuel air combustion design is concentrated on maximum power condition NO_x (usually high power condition efficiency and flame stabilization are of no problem). How the combustor functions from main fuel opening condition up to maximum condition without main fuel staging.

3.6. Main Fuel Injection Design

The main fuel injection design is very critical for this combustor. As shown in Fig. 3, main fuel injection is a combination of plain jet tube injector with co-axial flowing air. In technical literature this is often called co-flow air blast atomization. That definition is not completely correct.

When main fuel is just opened, main fuel injection pressure drop is very low, and without air, the atomization is extremely poor. The co-flowing air is the main factor to assist fuel atomization. So this is an air blast atomization mode. Then fuel injection pressure drop increases. At medium injection pressure drop, if there is no air, the atomization is still not very good, the co-flowing air is to assist atomization. At this period, it is air assist atomization. At high power condition, the main fuel injection pressure drop shall be very high. The atomization is good without co-flowing air and droplet size will be fine. With co-flowing air this co-flowing air functions to reduce the liquid jet-air relative velocity, thus air actually retards the atomization. Now it is in an air retard atomization mode. From an atomization point of view, it is not right to call such fuel injector as an air blast atomizer. The present authors suggest that fuel atomization is a combination of plain jet tube injector and co-flowing air. Under different conditions, fuel atomization can undergo different atomization mode.

From combustor design point of view, atomization is not the only one important factor to be consider. It is only true that under some conditions, when the fuel pressure drop is very low, droplet size is important. For the main fuel injection in the present design, when fuel pressure drop is high and at high power condition, fuel penetration, dispersion, and then fuel air mixing are more important than droplet size. Note that when droplet size is very fine, the fuel spray cannot outwardly spread a great distance. For this combustor design the main fuel injection possesses an angle relative to the module center. The injection angle relative to module center line is very important. At low power condition, the main fuel has little penetration. The main fuel mostly will be collapsed by pilot fuel combustion, which makes the flame stable and combustion efficiency good. At high power condition the main fuel penetration increases and is more important. With an angle relative to the center line, main fuel will also move radially outwards to meet the main air. With appropriate aerodynamic design, main fuel-air mixing is good to form main fuel lean direct mixing (LDM) combustion.

Because of such design idea this angle cannot be too small. If it is too small, such as lower than 10°, at maximum condition main fuel radially outwards penetration is insufficient. The main fuel - main air mixing is poor, maximum condition NO_x will be high, and combustion efficiency is not so good. If the angle is too large, such as higher than 20°, at low power condition, the main fuel will detach from pilot fuel combustion, and low power condition combustion efficiency will be poor. At maximum condition, the main fuel may have too much radial penetration, and main fuel over-penetrate through the main air, also causing poor mixing. Thus, it is not that the more penetration the better. Rather, the designer must seek an optimized penetration and dispersion. That is the key point. No less; no more. As shown in Fig. 3, which is only an example, this angle is 15°. Depending upon module dimensions, separation distances, main air swirling, the amount of co-flowing air (air liquid ratio), main fuel tube injector diameter, main fuel tube injector length (length over diameter ratio), and the distance between two neighboring modules, this angle shall be adjusted to provide maximal condition for best main fuel combustion.

As shown in Fig. 3, main fuel injection is not a hole. Rather, it is a section of small tubing. In this way this assures that the main fuel injection possesses the correct angle and radial dispersion. If it is only a simple hole, the fuel jet direction would be too much affected by manufacturing.

An optimal amount of co-flowing air must be inducted. If it is too little air, there will not be sufficient assistance for low power condition main fuel atomization. Also, there would not be sufficient assistance to optimize high power condition main fuel penetration. But if there is too much co-flowing air, that will increase pilot air module size, which also is not beneficial for low power condition main fuel combustion (thus, main fuel combustion is too lean). At high power condition there is no need to have the co-flowing air to excessively retard atomization.

Another important design variable is maximum condition main fuel injection pressure drop. Generally speaking, it is very high, such as values higher than 500 *psig*, which depend very much on fuel pump capability. (The next generation aero combustors actually will require very high pressure fuel pumps, with values such as 2000 *psia*.)

At maximum power condition the main fuel combustion and pilot combustion should possess the same equivalence ratio. Based on the maximum condition pilot air - main air division, and main fuel flow rate, the pilot fuel flow rate can be determined. After the maximum condition main fuel injection pressure drop is determined, the main fuel tube injector flow number is determined. Choose a number of main fuel tube injectors per one module to achieve the flow number for one main fuel tube injector. Because they are plain jet injections, ideally there shall be more injectors to provide more uniform main fuel circumferential distribution. Thus, there needs to be a compromise between the number of injectors and the tube injector diameter. Commonly used discharge coefficient for plain jet co-flow air blast atomizer cannot be used here because it would be a tube with a high length-to-diameter ratio, and there currently is no such data available. The designer must collect such injector discharge coefficient data via a single module tubular combustor test. For a first time design one might assume it will be 0.6. Then one would determine the tubing diameter, usually between $0.02 - 0.03$ *in*. The designer may choose a tube diameter, then determine the number of tube injectors on one module. The design process would need several steps of refinements. The designer would note that the circumferential arrangement of these co-flowing air tube injector combination may affect the liner exit radial temperature profile. They may, or may not be uniform distributed.

3.7. Main Air Module Design

Main fuel combustion air flow rate is total combustion air flow rate minus pilot combustion air portion. Main module air is main fuel combustion air minus main module cooling air, minus the main fuel co-flowing air. Note that main module cooling air flow is not perpendicular to the main module inner wall conical surface. The module air cooling flow is in the direction of the module center line axial to avoid local recirculation.

Main module dimensional design starts from main module AC_d, much the same way as the determination of pilot module AC_d. This is the main module required AC_d. The design of the main module inlet AC_d would be 1.8 times main module required AC_d. The main module inlet has two portions: non-swirling air, and swirling air. Non-swirling air fraction and main module swirling air strength are determined by an air flow pattern study. Previously, it was mentioned that the entire air flow pattern should have main air just passing by the pilot air recirculation. But for main fuel combustion at high power condition, the designer must evaluate the air flow pattern to achieve good main fuel main air mixing. Both the non-swirling air portion and main air swirling strength will be determined by a study of air flow pattern. In some cases, there may need to have non-swirling air on both the inner and outer sides of the main air module. To better manage the air flow pattern, adding an outer wall side non-swirling air is also a possible design choice. In this case, the main air swirler will possess a swirling strength 60°.

Main air module exit AC_d will be determined by Equ. 6. Exit geometrical area is equal to 1.03 times the exit AC_d. Exit location is determined by the requisite separation distance.

3.8. Single Module Tubular Combustor Cooling Design

Preliminary design of single module tubular combustor cannot be finished without a cooling design. For a CMC full annular liner, the single module tubular liner must also be CMC. That dictates that the mechanical design must be incorporated with aero thermal design, although a tubular combustor usually has no cooling issues. The cooling design is based on maximum power condition, with the same liner inlet pressure, the same liner inlet air temperature, the same liner FAR, and the same liner total pressure loss coefficient as the engine cycle parameters. Single tubular liners have the same air division as full annular combustors. Liner cooling is of a compound angle tangential inlet hole configuration design as shown in Fig. 6. It incorporates a compound angle, and there is an axial direction angle. Here D is liner diameter. For example, a liner diameter of 6 *in* with a liner wall thickness of 0.08 *in* (the wall thickness differs for a CMC tubular liner). The tangential hole edge shall be as close to the liner wall inner surface as possible. That proximity will be best for cooling effectiveness. Cooling air flows spirally downstream and compactly adheres onto the liner wall inner surface. For a machined tubular liner the distance from cooling hole center line to liner inner surface is just half of the hole diameter, which is

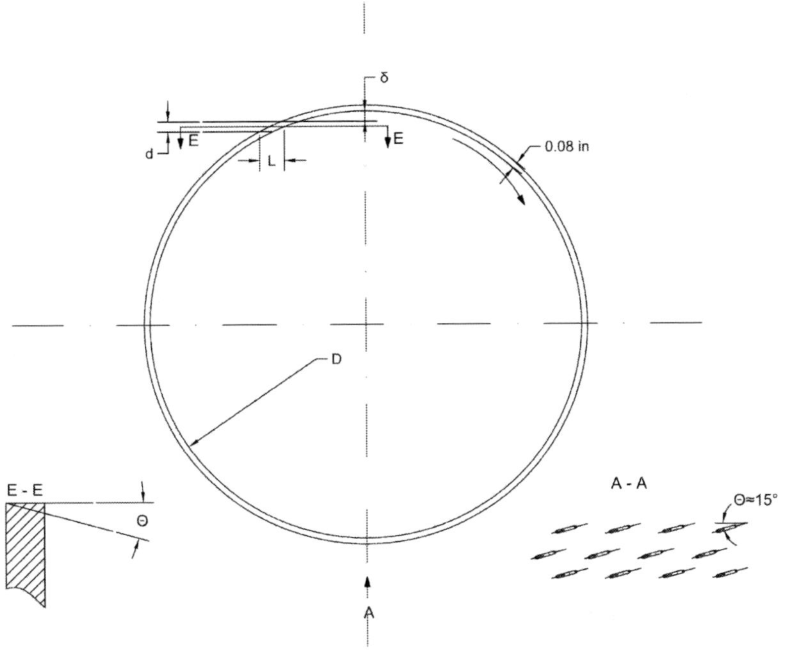

Figure 6. Cooling hole configuration for outer annular liner or tubular liner.

perfectly tangential. The hole diameter is 0.02 *in* (the hole diameter will differ for a CMC tubular liner). The axial direction angle depends upon axial spacing and circumferential spacing, which is shown in Fig. 7. For example, if axial spacing distance is 0.25 *in*, and suppose circumferentially there are 60 holes, the circumferential spacing distance is $3.1416 * 6/60 = 0.31416$ *in*. As shown in Fig. 7, the axial direction angle is $arctan(H/3 * S)$, which is 15°. The design rule is that the cooling air from the front flow cooling holes shall not impinge on the downstream row cooling jets. Note that for such a single module tubular combustor the cooling hole axial spacing distance is usually constant. But for any full annular liner, the axial spacing may be of an uneven design.

This cooling design has one additional advantage: the cooling hole length is rather long, much longer than the cooling air passage inside a Lamilloy or Transply (although this is not the major advantage over them). The hole length is defined by the hole center line length. As shown in Fig. 8 the hole length vs the distance from hole center line to liner inner surface for different liner diameters

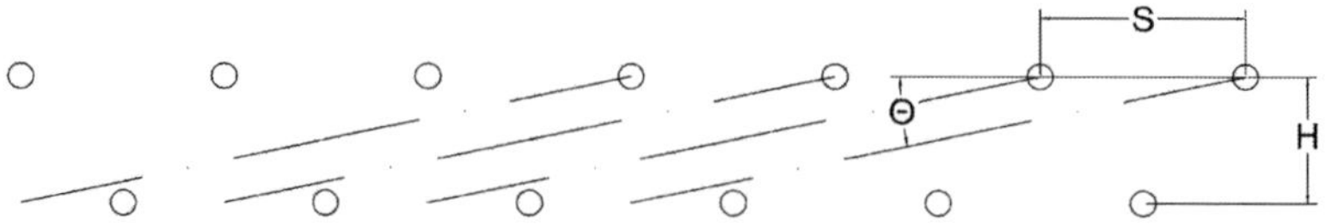

Figure 7. Design of the axial direction angle in compound angle tangential inlet cooling hole configuration.

are shown. Here, for a hole diameter of 0.02 *in* and liner wall thickness of 0.08 *in*, the liner diameter may be from 5 *in* to 30 *in*. For a 6 *in* diameter machined liner the cooling length can be 0.5 in long. Such cooling technology has been proven by combustion testing to be very effective. For low emissions combustors the cooling effectiveness can always be taken as unity.

In considering design choice for an optimal cooling air channel annular height, the average cooling air velocity within the annular channel at the dome's side could be 30 m/sec. Usually, for a single tubular combustor tested in a single module tubular combustor rig, the liner diameter is equal to the rig instrumentation section diameter. Thus, there will be no liner section contraction, and combustion gases will flow directly into the water cooled instrumentation section, while the instrumentation section jacket cooling water is discharged into an exhaust pipe to reduce the combustion gas temperature before reaching the exhaust control valve.

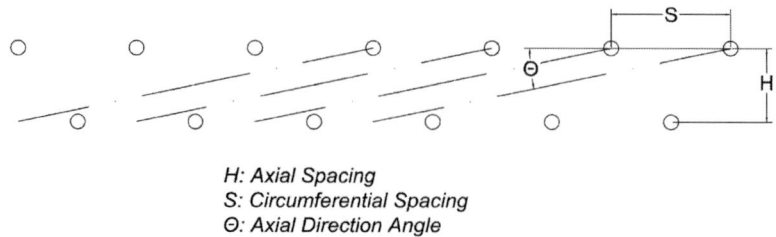

The prototype single module tubular combustor has no diffuser. So far this preliminary design has been finished. The development will run all kinds of tests: ground ignition testing, idle LBO testing, simulated high altitude ignition testing, 30% power condition heavy weather raining flame out testing, 4 cycle condition and maximum cruise condition performance testing, emissions measurements and efficiency testing under all power conditions, measurement of wall and dome temperatures, monitoring of injector pressure drop, monitoring of liner AC_d, monitoring of combustion instability, etc. There will be no liner exit temperature distribution measurement and no way to evaluate flame propagation between modules during ignition testing. The development of single

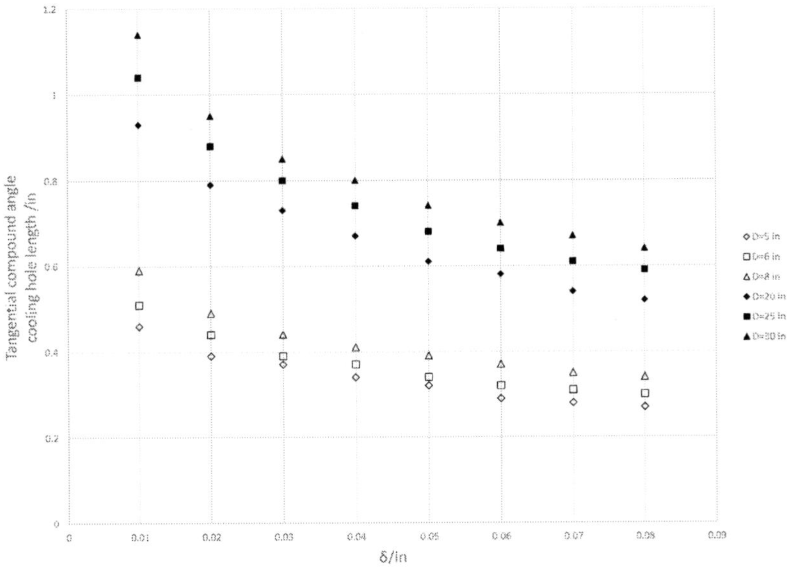

Figure 8. Tangential compound angle cooling hole length changed with the distance from hole centerline to liner inner surface for tubular and outer annular liner.

module tubular combustor will be discussed later in the "development" section of this book. After the single module tubular combustor development stage is finished, the preliminary design-development stage then is completed, and the next stage of combustor development will begin. The next section will discuss full annular liner design.

Chapter 4

Type One High Pressure Low Emissions Combustor (without Fuel Staging) Detail Design

This detail design may include a 90° sector combustor design and a full annular combustor design. Because in some cases a full annular combustor test rig cannot run up to maximum power condition, hopefully there will be a chance to run a 90° sector combustor test up to maximum power condition (only asking 25% of its maximum condition air flow rate). Even if the full annular combustor rig can be operated up to 100% power condition, a 90° sector combustor is still one step forward in the field of combustor development.

4.1. Full Annular Liner Design

For next generation aero combustors a full annular combustor design differs greatly from all previous conventional combustor design details. Now, the fuel-air module has been fully developed. That means such design details will require fewer alterations. Particularly, no change of the combustion organization design, no change in aerodynamic design, and no change of fuel injection design.

In preliminary design stage, the full annular cross sectional area combustor, the total number of fuel air modules and the air distribution between combustion air and cooling air, plus the air flow division between pilot combustion and main combustion, the fuel flow division between the pilot and main at idle condition and at maximum power condition, all have been determined. The next step is to define an optimized full annular liner contour.

At the combustor inlet, which is the compressor exit, there is an annular area. Its average diameter and annular height are known. At the turbine inlet, and also at the combustor exit, there is an annular area. Its average diameter and an annular height are known. Connecting these two average diameter points is a line that determines the liner average line which starts from the lower position gradually turning upward toward a higher position. But liner average diameter line is not just a straight line. At its beginning, from the diffuser inlet to the dome, it turns a small amount, say only 5 degrees (note that it is the average line of the diffuser). For the remaining portion it is the line from the dome to the liner exit, and the liner average line there will turn more, probably at dome side to 9 degrees or more. Then near the exit region, the turning angle diminishes. The diffuser length is 4 inches, and that is the axial position of the liner dome. This 4 inches are for the inlet diffuser (this is only for the new diffuser design). Based on this liner average diameter and the calculated liner cross sectional area, one can determine the outer annular liner and inner annular liner radial positions. Then one can assign 55% cooling air to be routed through outer annular channel, and 45% cooling air to be routed through the inner annular channel. The design of the outer annular cooling air channel induction location would feature a cooling air average velocity of 50 m/sec to determine the outer annular channel's annular height, and to determine outer air casing's inner surface position. Design of the inner annular cooling channel induction location would feature a cooling air average velocity of 40 m/sec to determine the inner air casing's outer surface position. Contour for the annular liner and the air casing are defined. Towards the liner exit there is a convergent section where the outer annular liner turns inward and where the inner annular liner turns outward, and is very much dependent upon the assembly of the liner into the air casing. There are several aspects to consider wherein the aero thermal design is closely related to mechanical design. The liner exit section is one aspect. The second aspect is the pilot module (including fuel injector) installation through the air casing and diffuser wall. The third aspect is how to connect the inlet diffuser to the air casing, and how the diffuser is in contact with the dome. The most important aspect of these questions is the CMC liner, its liner material, and the related mechanical design issues that must be well addressed before any aero-thermal design commences.

With a liner wall thickness of 0.08 *in* (different for a CMC liner), the liner wall material possibly could be Hastelloy X, or Haynes 188, or Haynes 230, or CMC. At this stage, for the liner cooling calculations, the liner material must be specified. For this same reason the air casing wall thickness and material also must be specified.

As previously mentioned, the fuel/air module has been aero thermally developed. In detailed design stage two things must be modified. One is to changeout the fuel injector from a single tubular combustor type, which is a tip-on-top configuration, to an engine combustor configuration without any modification of the essential design. Then we need to reposition all the air/fuel modules onto the dome to determine the optimal spacing between the two modules, and then evaluate the air flow pattern between the two neighboring modules. Because every main air module has its own air swirling clockwise (as viewed from downstream toward upstream). Between the two neighboring modules the air flow is in the opposite direction. There should not be any flow interference. If the spacing between two modules is insufficient, the air flow streams from two neighboring main air modules may interfere each other, and that is not acceptable. Also we will evaluate the fuel/air distribution in the areas between two modules. There should not be any main fuel overlapping in this area. Unless there is no other choice we will not change the total number of fuel/air modules (because that means another single fuel/air module design and tubular combustor development). The designer may move the dome center point radially outward to a larger average diameter position and so correspondingly change the liner average line position and the front side diffuser average line turning angle along with the liner average line turning angle. Such an alternation is very common. Note that any change should keep the liner cross sectional area unchanged, while also keeping the outer annular channel flow area and inner annular flow area unchanged. That change requires the air casing position be changed. Actually, such air flow pattern study should be completed firstly, during the preliminary redesign-development stage to avoid any problems.

4.2. Transient Operation Design

For combustion organization design, there is one thing left: transient operation design. For conventional combustor, between main fuel and pilot fuel there is a flow divider valve. At certain designed pilot fuel nozzle pressure drop (corresponding to certain operational condition), the flow divider valve opens, main

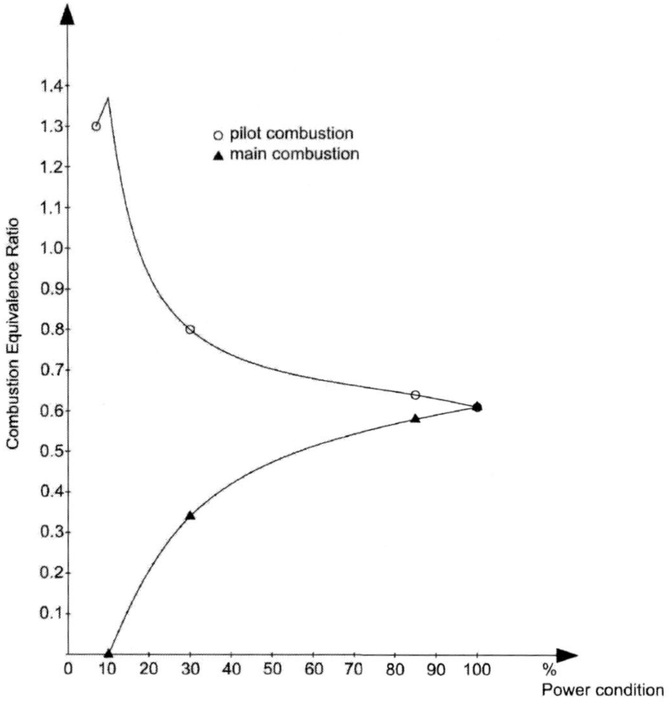

Figure 9. Combustion equivalence ratio changing with power condition.

fuel starts to work, this is called crack pressure. For this combustor, at idle condition main fuel shall not open to work, the crack pressure must be higher than the pilot fuel nozzle pressure drop at idle condition. For the present design that is higher than 180 *psig*. For maximum condition, the design shall use very high main fuel injection pressure drop. That means, suppose the design is still to use a flow divider valve, at maximum power condition, pilot fuel nozzle pressure drop is main fuel nozzle pressure drop plus crack pressure, or at least 680 *psig*. Then at maximum power condition, the required fuel split between main fuel and pilot fuel can not be achieved by such injector pressure drops, they are in contradiction. This is not desirable. The designer has no way to satisfy the requirement of fuel injection pressure drop, requirement of fuel split and flow divider valve crack pressure, at the same time. That determines a very important change from conventional combustor design: the simple flow divider valve can not be used. There shall be a separate control system to manage pilot fuel and main fuel division. This also means management of pilot fuel combustion equivalence ratio and management of main fuel air combustion equivalence ratio from main fuel opening condition up to maximum power condition. The desirable change of pilot fuel equivalence ratio and main fuel equivalence ratio with power condition, for this combustor, is shown in Fig. 9. Fig. 9 shows at idle condition, pilot fuel combustion equivalence ratio is 1.3. Main fuel equivalence ratio is zero. Very soon, at about 10% power condition, main fuel opens to work. At 30% power condition, pilot combustion equivalence ratio is 0.8, which is good for efficiency, NO_x and flame stabilization. At nearly 60% power condition, main fuel injections will detach from pilot fuel combustion to form its own combustion zone (main fuel equivalence ratio is more than 0.4). At 85% condition, pilot fuel equivalence ratio is 0.65. At 100% power condition, both combustion zone equivalence ratio is 0.625. Here the key element is from main fuel opening condition up to maximum condition, pilot fuel and main fuel equivalence ratio are both controlled by a special control system. As compare to type two high pressure low emissions combustor design (reported later, which has fuel staging), if there is no such kind fuel split control system, the present combustor will have higher NO_x and can not meet the low emissions requirement. Thus in order to take the advantage of simplicity without fuel staging, development of fuel split control system to replace the conventional flow divider valve is necessary. By the way, whenever aero combustor development entering a brand new era, very often there will be some new technology to replace the existing technology. Development of fuel split control system is one big step forward in aero combustor technology.

For high pressure low emissions combustor aero thermal design, there is one very special issue left, that is, without dilution air, how can the designer manage exit radial temperature profile? The present author suggests that from single tubular combustor test to full annular liner development, the total number of main fuel tube injectors in one fuel air module shall not be changed, but the circumferential arrangement of main fuel injectors can be modified, or even from uniform arrangement changed to non-uniform arrangement, as shown in Fig. 10. This is the way to manage full annular liner exit radial temperature profile. Such development shall be incorporated with high pressure sector combustor test. It costs too much to adjust exit radial profile in full annular combustor test. If change of main fuel tube injector circumferential arrangement can not solve the problem, then another design approach is to have some air holes directly

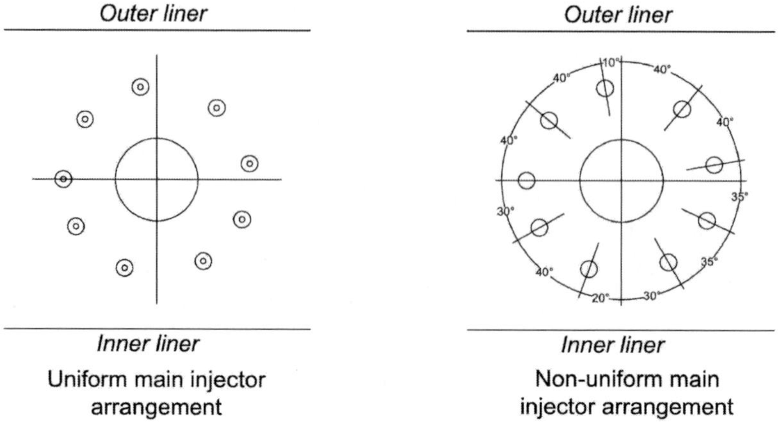

Figure 10. Different main injector arrangement for liner exit radial temperature profile.

through dome, but which can only be limitedly used.

4.3. Full Annular Combustor Cooling Design

First thing to be noticed is that, it is whole combustor cooling design and calculation, not just liner cooling design and calculation. It will be noticed that although liner cooling design and calculation is the major purpose, but determination of liner temperature is related to the determination of air casing temperature. That is the reason, the designer must take the whole combustor as a heat transfer system, not just the liner, to determine the system temperature.

In single fuel air module tubular combustor design and development, there is no cooling calculation, usually it is of no problem. For annular liner detail design, air cooling calculation is a must.

4.3.1. Outer Liner Cooling Design

Outer liner cooling configuration is the same as tubular liner, as shown in Fig. 6 and Fig. 7. From Fig. 8, it can be seen that the cooling hole length is very long. Because the liner is of large size, usually it is sheet metal rolled and welded, the distance from cooling hole center line to liner inner surface shall be more than half of the hole diameter, depending on liner roundness, to avoid laser drilling blind holes, it may be 0.03 *in*, from Fig. 8, for liner diameter 25 *in*, the hole length is 0.8 *in*. For CMC liner, there shall be very special design considerations.

4.3.2. Non-Uniform Cooling Hole Axial Spacing

In order to save cooling air and also to help the liner temperature distribution along axial distance to be more uniform, cooling hole axial spacing distance is not constant. At near dome side, combustion temperature is gradually going up, thus here the axial spacing can be greater. As getting to 4 *in* from the dome position, the combustion is going to be finished, so in this region the axial spacing distance is minimum. Gradually moving towards exit, the axial spacing distance can be higher again. This is a long time observation result from combustion test, the present author noticed that the liner wall temperature distribution along axial direction is always looking like a hill. For a good combustor, the hottest point is always about 4 *in* from the dome. Non-uniform axial spacing design not only can save cooling air, but also may provide a more uniform wall temperature axial distribution which will reduce thermal stress.

Near outer liner exit region the outer liner diameter is getting smaller. In this region, the design may keep the circumferential hole number constant (for manufacturing consideration), and further enlarge the axial spacing. The present author noticed with such kind cooling design, once the compact cooling air layer has been formed, it can protect the liner for a rather long distance, even with wide axial spacing.

For such annular liner the cooling air division between outer liner cooling and inner liner cooling is 55% vs. 45%. Often the ratio between outer liner surface area and inner liner surface area is 60% vs. 40%. As inner liner cooling is a little bit more difficult than outer liner, thus inner liner cooling has more priority.

4.3.3. Inner Liner Cooling Design

Inner liner cooling configuration is very different from outer liner cooling. Because if using the same cooling configuration, the cooling air will flow away from the wall surface to mix with combustion gas, which is very bad for cooling effectiveness.

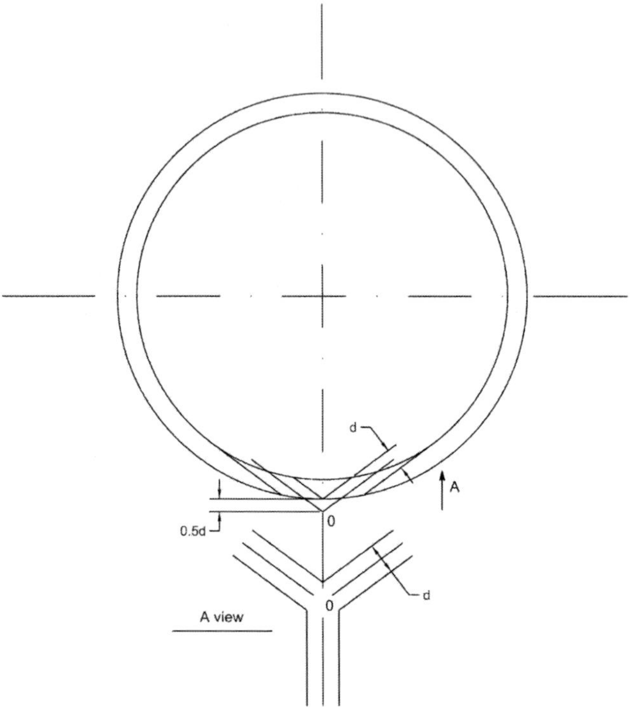

Figure 11. Cooling hole configuration for inner liner.

Inner liner cooling configuration design is based on one concept that two compound angle tangential flowing air jets impinged at wall surface will form a pure axial direction cooling air jet. The configuration is shown in Fig. 11. Both impinging jets have an axial direction angle. This axial direction angle must be larger than a certain value, as shown in Fig. 12, to avoid the impinged jets to form a reversed flow. This minimum axial direction angle depends on liner diameter. Of course, no need to have axial direction angle greater than the necessary one. Again, for CMC inner liner, there will be some special design considerations.

In conventional combustor there is machined ring cooling, which has one advantage that cooling air is flowing axially. But the machined ring has a lip,

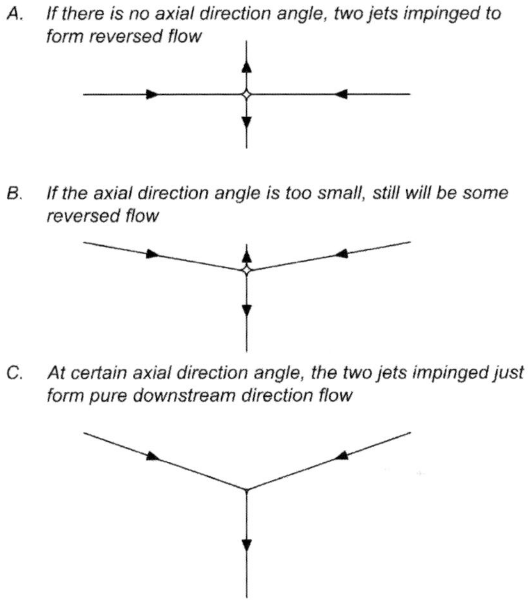
A. If there is no axial direction angle, two jets impinged to form reversed flow
B. If the axial direction angle is too small, still will be some reversed flow
C. At certain axial direction angle, the two jets impinged just form pure downstream direction flow

Figure 12. For inner liner cooling hole there must be axial direction angle.

which is life limiting factor. Now the new inner liner cooling technology has axial flowing cooling air without a life limiting factor.

For inner annular liner cooling, again the axial spacing distance is not constant. The circumferential spacing shall be smaller than that in outer annular liner. The single hole diameter shall also be smaller than that in outer liner. In inner annular liner there is a turning point where the liner is bending outer wards. At this location, there can be a number of small holes to make a purely axial cooling air flow sticking on the exit region of inner liner wall.

This cooling configuration design is special. But it brings some challenge to laser drilling. There are two ways to do the laser drilling. First, at one circumferential location, laser drilling one direction cooling hole, then immediately turning to have another direction hole laser drilled. Such method will increase drilling time. Second, at one location laser drilling one direction hole, then turn circumferentially moving to another circumferential location, drilling the same direction hole, and so on. After finishing all the one direction holes, then turning to laser drill another direction holes. This way can save time, but good alignment will be an issue, as turning 360°. coming back to the initial location, not easy to have good alignment with the first drilled hole. For CMC liner, laser drilling tiny small holes is a challenge. And thermal barrier coating maybe necessary.

For low emission combustor, usually no need to use thermal barrier. But some time it is worthwhile to pay several thousand dollars cost to have liner life extension several thousand hours.

4.4. Cooling Calculation

Liner cooling calculation is based on maximum power condition or at 120°F hot day condition. As it will be seen later, in order to calculate liner wall temperature, it is necessary to simultaneously solve both liner heat balance equation and air casing heat balance equation. At the same time, also getting air casing wall temperature determined, which is good for air casing stress analysis. Notice for CMC liner, the cooling calculation method is same. But material properties are very special.

Very often in some gas turbine combustion book, the liner cooling calculation will calculate the whole liner axial temperature distribution. But for combustor development, it is important to have hottest wall temperature defined, very often it is a region nearly 4 *in* distance from the dome. The calculated liner wall temperature is liner outer surface temperature (T_{w2}). This temperature is going to be compared with the same location T_{w2} measured during combustor testing (which is an averaged value of circumferentially multi-point thermocouple measurements).

There are two types combustor liner and air casing cooling calculation:

- outer annular liner and outer air casing wall temperature calculation
- inner annular liner and inner air casing wall temperature calculation

The basic principle is the same for both types.

The next generation low emissions combustor cooling calculation method is very different from previous conventional combustor cooling calculation. The cooling calculation needs to be updated. The present author will outline those updated:

- Because the combustion is much cleaner, there is no luminous flame, only gaseous phase radiation. Gaseous phase radiation heat transfer may be calculated rather accurately by present author's research, which will be reported later.

- Liner cooling air effectiveness is very good, for outer liner, it is 100%. For inner liner it may be between 90% and 100%.
- Liner convective heat transfer is not from hot gas to wall, it is from hot wall to cooling air layer. Because there is a compact cooling air layer on the wall surface. This layer temperature is lower than hot wall temperature.
- The net heat transfer from liner wall to the air in the annular channel is radiation heat flux minus convective transfer heat flux (which is from wall to cooling air) minus the heat absorbed by the air passing through the cooling hole (calculated for one cooling hole heat transfer, then times hole number per unit area).
- air casing wall temperature is not equal to air temperature T_{31}. Difference can be more than 100°F. It is necessary to have heat balance equation for liner and heat balance equation for air casing, to calculate not only liner temperature (T_{w1}, T_{w2}) but also air casing wall temperature at the same time, as shown in Fig. 13:
- For liner wall, K_{1-2} is not equal to R_1–C_1 , it should be:

$$K_{1-2} = R_1 – C_1 - Q_{ah} \tag{7}$$

where

K_{1-2} conduction heat flux through wall

R_1 internal radiation heat flux from gas to wall

C_1 internal convective heat flux from wall to cooling air

Q_{ah} heat flux carried by cooling air passing through cooling hole

But the equation to determine $(T_{w1}–T_{w2})$ is

$$R_1 – C_1 - 0.5 * Q_{ah} = (\lambda_w/t_w) * (T_{w1} – T_{w2}) \tag{8}$$

where

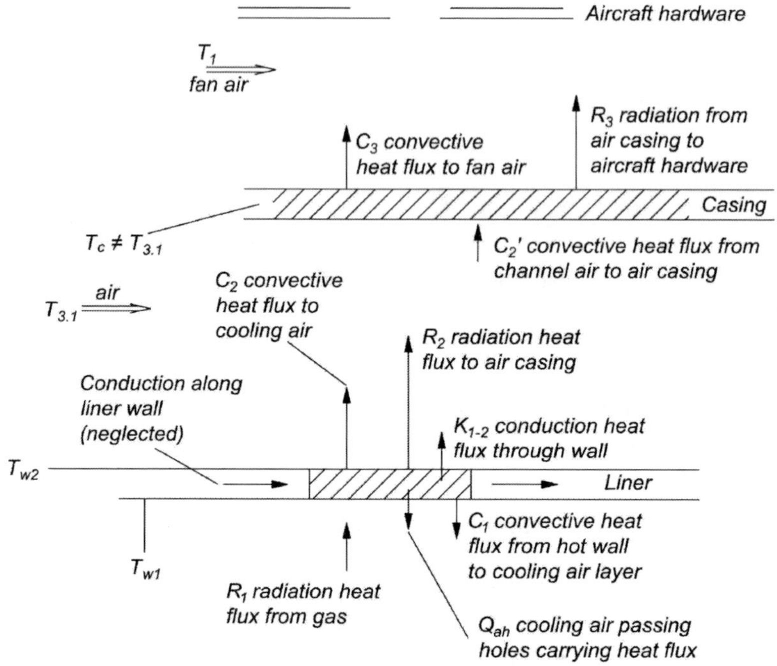

Figure 13. Basic heat-transfer processes.

λ_w thermal conductivity of the wall material

t_w liner wall thickness

Notice the λ_w value for CMC is very different from metal. Liner wall thickness for CMC liner is also different from metallic liner.

- Cooling air temperature out of effusion holes is not equal to T_{31}, after absorption of heat, the air temperature is increased.
- The heat balance equations for liner and for air casing shall be simultaneously solved by numerical method.
- For air casing heat balance equation, no problem to calculate the radiation from liner to air casing, no problem to calculate the convective heat transfer from channel air to air casing. To calculate air casing convective heat transfer to fan air, it needs to know fan air flow velocity. To calculate air casing radiation to engine nacelle, needs to know the nacelle wall

temperature. This is something not so easy to determine. This radiation heat transfer may be neglected, that will give a little higher calculated air casing temperature (truly little), which is on safe side for liner wall temperature calculation.

For inner liner and inner casing temperature calculation, the difference is that the inner air casing is having natural convective heat transfer as there is no flowing fan air.

4.5. Combustor Inlet Diffuser Design

What are the main features of the flow field at combustor inlet?

Combustor inlet flow field has relatively high Mach number and significant circumferential non-uniformity, with an asymmetrical, peaked inlet radial velocity profile. Also such features have appreciable variation with changes in power condition.

In conventional combustor, what kind of diffuser outlet flow field is desirable?

Diffuser outlet flow field shall be low velocity, some dynamic head is recovered to static pressure. Also most of the flow distortion has been "neutralized". Diffuser outlet air shall smoothly flow forward to three directions, outer annular channel, central dome to liner, inner annular channel.

Present author has identified that combustor inlet diffuser has three functions (not just diffuse the flow to lower velocity):

- reduce the velocity to recover dynamic head to static pressure
- function like a reservoir, to smooth out flow distortions
- function as an air distributor, provide air flow to three different passages

Why, long time ago, combustor inlet diffuser design was changed from aerodynamic diffuser (or faired diffuser) to dump diffuser?

Aerodynamic diffuser has two sub-diffusers (one to outer channel, one to inner channel) with their rather small size annular height between two sheet metal walls and one central flow to dome after the snout. Due to manufacturing tolerances, different thermal expansions between liner and air casing, the effective flow area of these three branched flows will change a lot. The result is that the air flow distribution for these three streams will change significantly with change of operational conditions, or change from one combustor to another combustor, then the liner exit temperature profile will change correspondingly. This is the major reason that aerodynamic diffuser configuration was changed to dump diffuser.

Compare the total pressure loss in dump diffuser and that in aerodynamic diffuser, it is not very much different. As the dumping total pressure loss is higher than the flow separation loss in aerodynamic diffuser, while the friction loss in aerodynamic diffuser is more than that in dump diffuser. Dump diffuser has one advantage, the liner exit temperature distribution is less sensitive to inlet flow field. Also it helps to reduce the liner total length. Thus dump diffuser has become the predominant configuration for combustor inlet diffuser. But as compressor pressure ratio is getting higher, combustor inlet Mach number is also getting higher, dump diffuser total pressure loss is more. This is undesirable. As diffuser pressure loss and liner total pressure loss are of different nature. Any total pressure loss in diffuser makes no contribution to combustion.

Another situation has happened with the development of next generation aero combustor, that is, combustion air fraction in the total air flow has greatly increased, for example 75% and more. That means, 75% air will go through dome into liner. That significantly changes the flow distribution between three branches. In conventional combustor, very often the distribution is more or less evenly (not exactly one third for each branch). For next generation aero combustors, if combustion air is 75% or more, only 25% or less air will go to outer and inner channel. That may bring very big change in diffuser design as shown in Fig. 2. It is air bleeding diffuser proposed by the present author.

This diffuser will stretch directly to touch dome, there may be very small gap, not possible and no need to totally seal this gap. In any case, the diffuser is bleeding air out. The fuel injector-pilot modules are installed through air casing openings and openings on diffuser outer wall. There is a baffle on every fuel injector stem to cover the openings on diffuser outer wall, again not possible to totally cover the openings.

For diffuser air bleeding, the narrow width slots are better than circular holes. The present design proposes narrow width slots to have air bleeding out. There may be two bleeding slots on each side.

Using air bleeding to improve diffuser performance is a well known approach for long time. 40 years ago, the research on hybrid diffuser, as shown in Fig. 14, from Adkins, Matharu, and Yost (1981), did show some good results. But two things prevented this idea from really becoming combustor component.

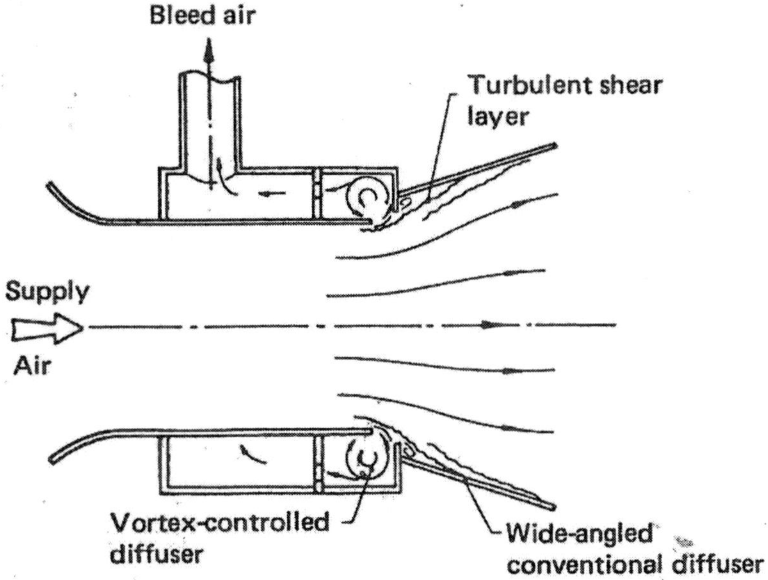

Figure 14. Hybrid diffuser arrangement, Adkins, Matharu, and Yost (1981).

They are:

a It is difficult to be integrated on to engine combustor. Notice at combustor inlet, it is also compressor outlet, space is limited.

b What can the bled air be used? There was no appropriate application to use this air.

The present author has learned lessons from such hybrid diffuser design.

The present author also learned from another very old paper (McLellan and Nichols 1942). As shown in Fig. 15, the tested conical diffuser has a very large expansion angle, half angle is 14°. Normally maximum expansion angle without flow separation would be 8°. This diffuser has a screen of 40 *eyes/in* at 2/3 inlet diameter, upstream of the screen there is no flow separation. Notice the flow resistance is inside the diffuser. This result showed that diffuser in combination with a flow resistance is beneficial to prevent flow stalling, then the expansion angle may be enlarged to reduce the diffuser length, the total pressure loss caused by the screen is less than the total pressure loss saved by the reduction of diffuser length.

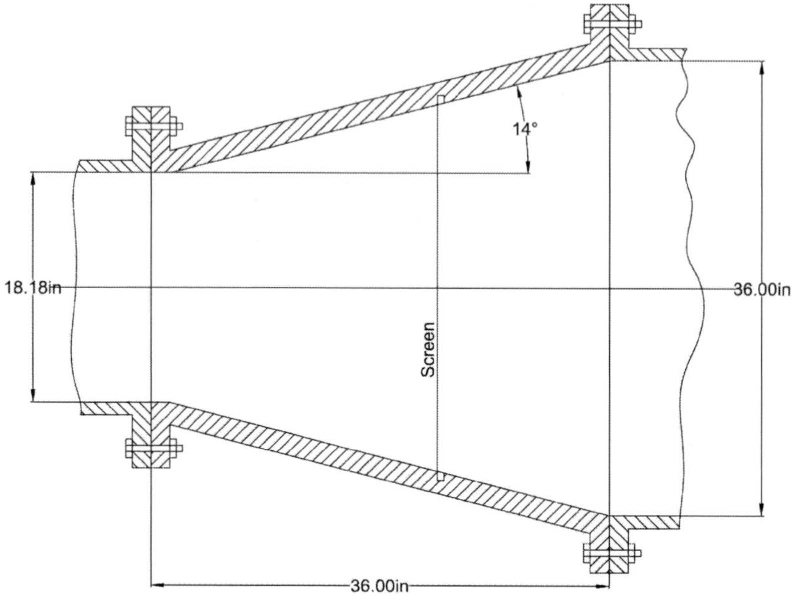

Figure 15. Large expansion angle conical diffuser with screen.

Combining these two research results together, the present author's design of air bleeding diffuser is shown in Fig. 16. Only diffuser outer wall is shown. The idea is to use the dome as flow resistance, then using air bleeding out of diffuser to have larger expansion angle without severe flow separation. The advantage is to have diffuser total pressure loss basically the same as dump diffuser, but diffuser length is reduced to 4 inches. The bled air is used for liner cooling.

The diffuser has two bleeding slots on each side. As shown in Fig. 16, the first one, symbolled as **A**, is at the beginning of the diffuser. It is a circumferentially non continuous slot. The half expansion angle is about 8°. Two inches downstream from the inlet, there is another bleeding slot, symbolled as **B**. This one is also not circumferentially continuous. The half expansion angle is about 8°. The diffuser outer wall is cast with the air casing as one piece. There are 8 thin struts to connect the diffuser outer wall with outer air casing. The 8 thin struts support the diffuser outer wall. On diffuser outer wall, the total bleeding slots have geometrical area twice of the geometrical area of outer liner cooling holes. Correspondingly on diffuser inner wall the two slots have geometrical

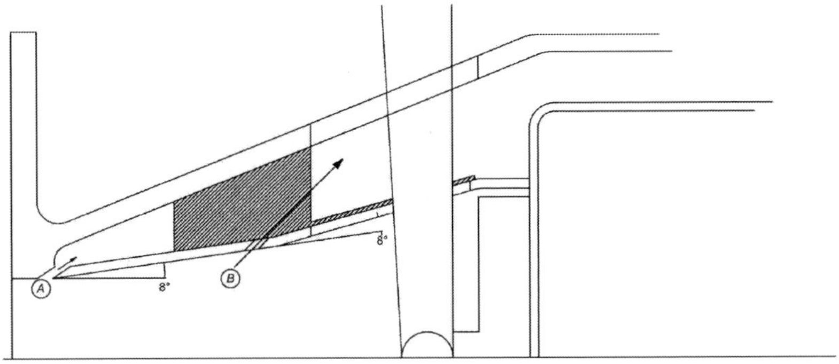

Figure 16. Air bleeding diffuser configuration (only outer diffuser wall is shown).

area twice of the inner liner cooling holes geometrical area. To summarize, two bleeding slots on each side, on each side the bleeding slots have geometrical area twice of the corresponding liner cooling holes total geometrical area. Liner air holes are still the metering device in the flow passage. Thus most pressure drop happens at liner cooling air holes, not at bleeding slots. There are two turnings on each diffuser wall side, each time turning half expansion angle is about 8°, each diffuser wall is supported by 8 thin struts. Diffuser inner wall design is much simpler than diffuser outer wall design, as there is no fuel nozzle passing through. The total diffuser length is 4 inches. Based on this limitation to design the diffuser half expansion angle on each side.

Chapter 5

Type Two High Pressure Low Emissions Combustor (with Fuel Staging) Design

5.1. Fuel Air Module

From previous study, it is clearly found that the pilot fuel air combustion organized by one axial air swirler and one simple pressure swirl atomizer, has good performance at low power conditions. It also has very good NO_X at high power conditions. The explanation is as follows. At high power conditions, air pressure and air temperature are high, fuel injection pressure drop is high, fuel atomization is good, drop size is fine, and fuel spray evaporation is fast. For NO_X, if the combustion is lean, then NO_X formation mainly depends on fuel air mixing. In order to have good mixing, fuel and air shall have close contact and small scale mixing. Such configuration provides fuel air close contact. Pilot module fuel and air are small scale mixing, at high power condition, it is close to premixing situation. For combustion flame zone, it will not distinguish a well mixed mixture of fuel vapor and air is from a premixing module, or it is well mixed after leaving the module during the process before reaching the flame zone. That is the reason such fuel air combustion organization has good NO_X at high power condition. This has been proven by combustion test.

Based on this finding, the present author has formed an idea: just use combination of one axial air swirler and one simple pressure swirl nozzle to design

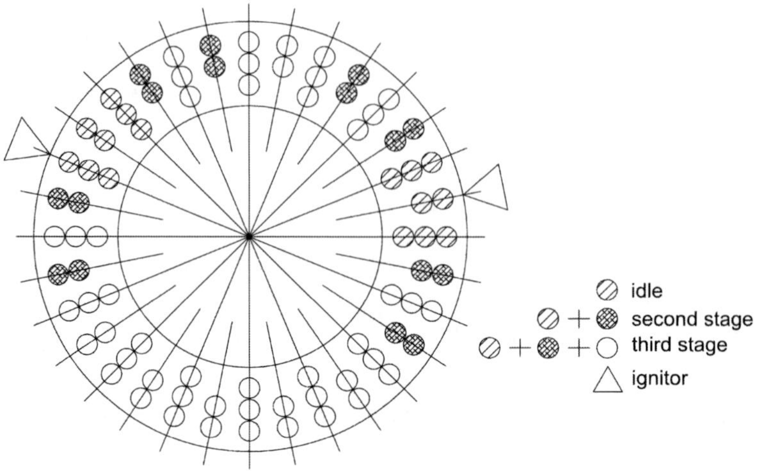

Figure 17. On dome fuel-air module arrangement.

an alternative high pressure low emissions civil aero combustor.

The arrangement of fuel air modules on liner dome is shown in Fig. 17, from Chin and Suo (2018). The single fuel air module configuration is shown in Fig. 18, also from Chin and Suo (2018), but with some modification. The modification is that, as the air swirling is all the same direction, in the neighboring area between two modules, the air flow direction is opposite, there may have flow interference. For some modules, there are four sides facing such problem, or three sides facing such problem. Thus in Fig 18, at module exit, there is a non-swirling air flow at module outer wall, to reduce such flow interference. Such non-swirling exit flow is only a small portion of pilot module flow but it will be useful. Sketch of a cluster of three fuel nozzles is shown in Fig. 19 (another type is a cluster of two nozzles). A typical opening on combustor air casing for the cluster fuel nozzles to pass through is shown in Fig. 20. The idea is from GE LM 6000 aero derivative gas turbine low emissions combustor. Where there are 75 premixing modules, arranged in three rows. They are all of same modules, and same fuel (natural gas) injectors. In present design they are not premixing modules, only fuel nozzles are passing through air casing in clusters, air modules are staying on dome. In this way the opening on air casing can be relatively small. There will be no difficulty for fuel nozzle installation.

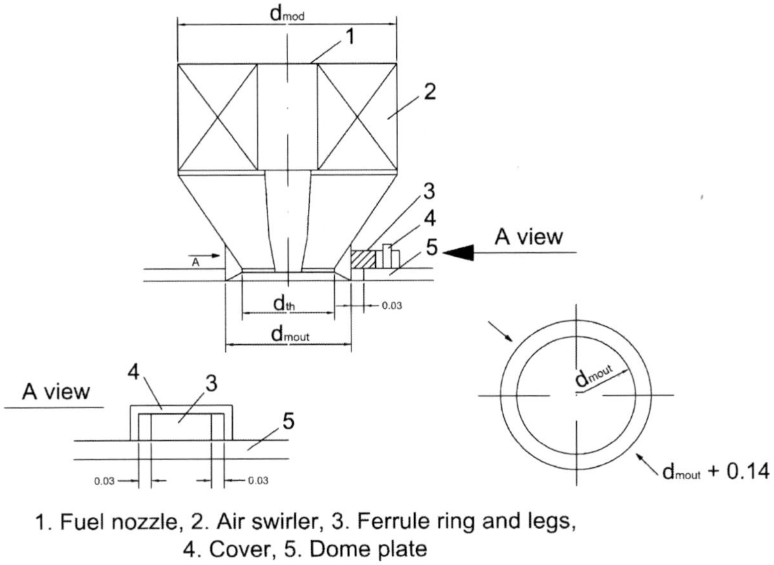

Figure 18. Fuel-air module.

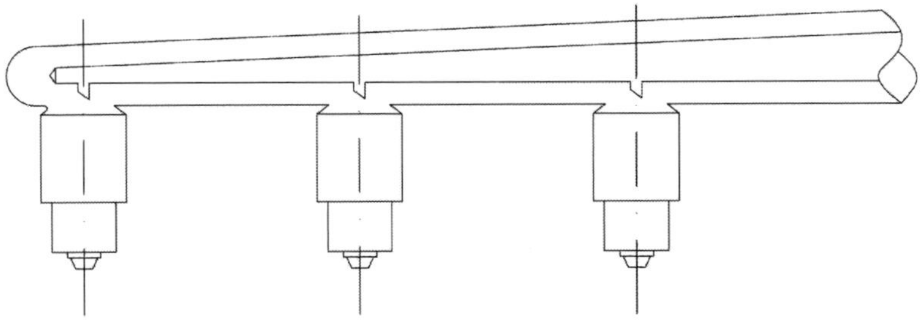

Figure 19. Cluster of 3 nozzles.

As shown in Fig. 17, the total number of fuel nozzles, N, is a critical design parameter. N is determined by the following equation:

$$3.1416 * [0.5 * (D_{in} + D_{out})] = N * (d_{mod} + 0.5)/2.5 \tag{9}$$

where:

D_{in} inner liner diameter, in

D_{out} outer liner diameter, in

d_{mod} single fuel-air module inlet diameter, in

N total number of fuel air modules

This equation simply means the total number of fuel air modules is determined by the middle row of modules circumferentially with 0.5 *in* spacing between two neighboring modules. The total number of modules is preferred to be 80, 72, 64, 56, etc., for the reason that it is convenient to run a 90°. sector combustor test.

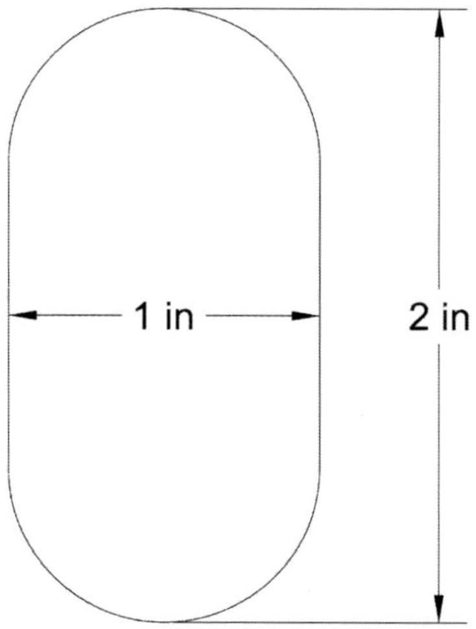

Figure 20. Opening on air casing for cluster fuel nozzles.

Radially, it is necessary to have 0.3 inch spacing between the module inlet and inner or outer liner wall, then radially there must be 0.2 *in* between two neighboring modules. The spacing between radial modules or between circumferential modules are not much. Because the fuel air module exit diameter is less than module inlet diameter, so the distance from module exit to liner wall and the spacing between two module exits are more than the above values. The radial arrangement of the two fuel nozzle cluster may not be at the same radial position as the three fuel nozzle cluster. Based on sector combustor test result, for exit radial temperature profile, the radial position of two fuel nozzle cluster (and their air modules) may be adjusted a little bit.

The design feature of such arrangement is that, there are three rows of fuel air modules, for example, as shown in Fig. 17, from outer liner to inner liner, they are 32, 32, 16 respectively. They are totally identical fuel nozzles and air modules of the same configuration and same size. All air alone modules are staying on dome, they do not pass through the openings on air casing, only the fuel nozzles are passing these openings as shown in Fig. 20. But the fuel nozzles are not passing the openings individually, they are passing in groups, as a cluster of three fuel nozzles or two fuel nozzles in one stem. As the clustered fuel nozzle has a size similar to a single nozzle, the opening on air casing is not big, only an ellipse about $1 * 2$ *in*. Because all air modules are staying on dome, it is difficult for a cluster of three nozzles exactly to fit the three central holes of three air modules. Because of this reason, some air modules have floating design as shown in Fig. 18. For the middle row air module, no need to have floating design, only the inner and outer row air modules have the floating design.

Design of single pressure swirl atomizer and single air module are the same as previously reported pilot fuel nozzle and pilot air module design. For this combustor, there needs fuel staging design.

5.2. Fuel Staging

Fuel staging design is concentrated on one question, whether it is possible to have just three stages and avoid to have four stages? The answer is, it is possible with high fuel pump pressure capability. The arrangement makes use of one fact that between 30% power condition and 85% power condition there is no emissions requirement.

For fuel staging design, let us introduce one concept: combustion enrichment factor. It is the real combustion zone FAR over liner total FAR. For example, for fuel air module design as shown in Fig. 3, suppose, at idle condition liner FAR is 0.013, all fuel concentrated in pilot combustion zone. The pilot combustion zone uses 14.7% air, thus although the liner FAR is very lean, but the combustion zone FAR is $0.013/0.147 = 0.0884$, it is equivalence 1.3. The combustion enrichment factor is $1/0.147 = 6.8$.

Combustion enrichment factor for fuel staging, is total number of air modules / working module number (this is only true when the air modules are all of the same AC_d). Or it is *total combustion air* AC_d */ working air module* AC_d.

Now design idle condition working fuel air module combustion equivalence ratio 1.3 to determine how many modules working during idle condition. For 30% power condition, the working module combustion equivalence ratio shall be at least higher than 0.7 (this specific fuel stage design is not suitable if 30% power condition liner FAR is lower than 0.016). It will have good flame stabilization and good efficiency.

Liner exit circumferential temperature distribution is non-uniform during staging, that is a drawback of this design. There needs some analysis to make sure the cyclic temperature fluctuation does not coincide with turbine natural resonance frequency.

In this specific design fuel staging is as follow:

- First stage - idle condition, there are 16 working fuel nozzles, fuel nozzle pressure drop 130 *psig*, just before opening of second stage (20% power condition), fuel nozzle pressure drop is 500 *psig*, just after second stage opening, fuel nozzle pressure drop is 100 *psig*.

- Second stage - from 20% to 50% power condition. There are 32 working fuel modules. At 30% power condition fuel nozzle pressure drop is 235 *psig*. Just before opening of third stage, nozzle pressure drop is 700 *psig*, just after opening of third stage, nozzle pressure drop is 120 *psig*.

- Third stage - From 50% power condition to maximum condition, there are all fuel nozzles working. At 85% power condition, nozzle pressure drop 340 *psig*. At 100% power condition, nozzle pressure drop 500 *psig*. During the whole operation process, minimum nozzle pressure drop is 100 *psig*, maximum nozzle pressure drop is 700 *psig*.

Fuel staging control can be by simple on-off valve.

5.3. Comparison of Two Types Low Emissions Combustor Design

The low emissions combustor design shown in Fig. 17 and Fig. 18, is expected to have better emissions than the design shown in Fig. 3, if the type one combustor design does not have fuel split control system. Because the main fuel and main air mixing design in Fig. 3 is not small scale mixing at high power

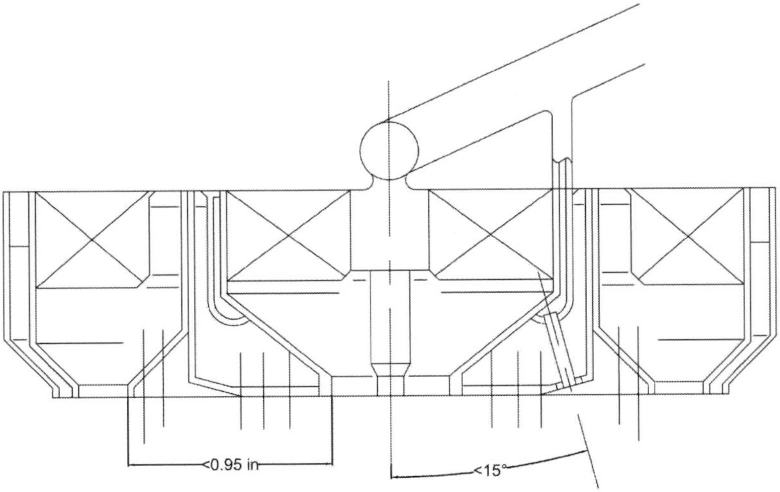

Figure 21. Fuel air module for extra high FAR combustor.

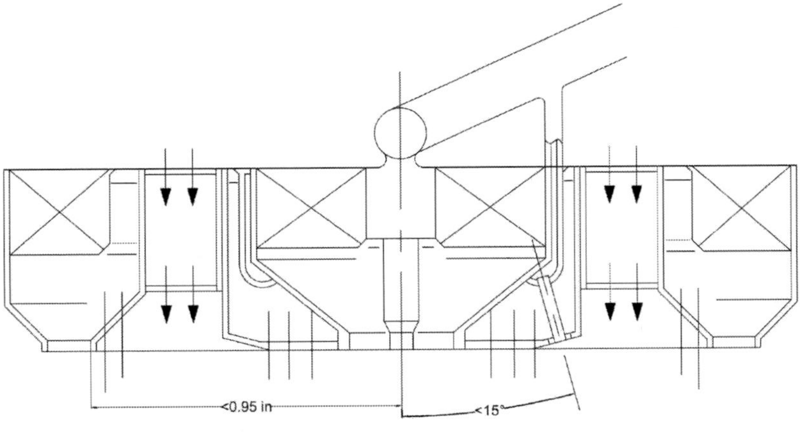

Figure 22. Fuel air module for small engine high FAR combustor.

conditions. The type two design is expected to have better high altitude ignition. Because the fuel air mixture for ignition is rather close to the ignitor. But the design shown in Fig. 3 has its merits. First, there is no need to have fuel staging. Second, if there is fuel split control system, then at high power condition, the combustor can always have optimized pilot fuel-main fuel split to have optimized emissions, under such situation, its emissions will not be poor. Third, for an engine institution, very often development of civil combustor and development of military combustor are both to be done in one organization. The design shown in Fig. 3 and Fig. 21, 22 are very similar, that is one fuel air combustion organization design suitable for both types of combustor. That means, the development of technology can be beneficial for both, such as technology for main fuel injection and co-flowing air combination can be used for both civil and military combustors.

Chapter 6

Design of High FAR Combustor

For military engine combustor, which differ greatly from civil combustor, there is no 30% power condition, nor a 85% power condition. There is no maximum cruise condition. For a military combustor there are ground idle condition, maximum power condition and cruise conditions at different altitudes and different Mach numbers. There is also altitude idle condition. Particularly there is a low altitude low Mach number penetration dash condition. In this condition the combustor inlet pressure may be higher than take-off conditions, and the inlet air temperature very close to take-off conditions, and FAR is only a little bit lower than take-off conditions. Cooling calculations may need to be used to evaluate for this condition.

Don Bahr reported that there are two major problems for high temperature rise (which is the same meaning as high FAR) combustor design, (Bahr 1987). These two major problems are idle LBO and liner cooling. Actually, these two problems are truly big issues, but also there are also other issues that must be considered by the designer, see Chin (2019).

Firstly, the idle LBO issue. The present authors use nearly the same fuel/air module configuration as shown in Fig. 3, but a little different design approach is used to solve the idle LBO problem. For a high FAR combustor, its idle condition FAR is higher than that of a civil aero combustor. That brings to mind a problem. If the designer still uses the design approach, at idle condition only pilot fuel is working from idle FAR down to flame out FAR. At flame out the pilot nozzle pressure drop may be less than 10% of idle condition pilot nozzle pressure drop. That is harmful for LBO. The design approach shall be changed; that is, at idle condition the main fuel is open to work. To design for idle condition, the fuel flow division between pilot nozzle and main injector is a design choice – for example, say, 70% metering through pilot nozzles and 30% metering through main injectors. In such a way, when the fuel is throttled the main fuel flow firstly is decreasing, and during this period the fuel flow decreasing is mainly achieved by diminishing the main fuel delivery, but the pilot fuel nozzle pressure drop only decrease a little. Then the main fuel is closed down, after which the pilot fuel starts to decline alone and pilot fuel nozzle pressure drop is going down proportional to fuel flow rate squared. At idle condition, a 70% idle fuel metering through the pilot fuel nozzles will form a stoichiometric pilot fuel air combustion. So, if we determine how much fuel is metered through the pilot fuel nozzles at idle condition (say, 70%), then we can design the pilot fuel/air combustion stoichiometric ratio to determine pilot air flow rate at idle condition.

Because at idle condition the main fuel is open to work and the main fuel combustion FAR at maximum condition is not lean but stoichiometric, that brings to mind the need for several changes of the fuel air module as compared to Fig. 3:

a The separation distance between the pilot module exit diameter and the main air module exit inner diameter is less than 1 *in*, while in Fig. 3 it is 1.0 *in*.

b The main module non-swirling air is divided into two portions. The inner portion is reduced to less than 20% of main module air flow. There is another non-swirling air located along the main swirler wall outer side which prevents possible over-penetration of main fuel injection. These two portions together make up one third of the main module air flow. It is also appropriate to drill holes directly through the dome to replace the outer non-swirling air as the main fuel is working at idle condition, and thus the air flow pattern may differ from Fig. 4.

c The main fuel injection angle is less than 15°, while in Fig. 3, it is 15°.

d In between two modules, in order to reduce the interference between two main module air, there may be a number of small holes directly from dome.

This combustor has a flow divider valve between pilot fuel line and main fuel line. That means the design of pilot fuel nozzle FN and pilot nozzle pressure drop at idle condition are related to the design of the main fuel injection pressure at maximum condition. At all operational conditions main fuel injector pressure drop is directly related to pilot fuel nozzle pressure drop via the flow divider valve crack pressure. Note that such a scenario is very different from the combustion organization design for a type one high pressure low emissions combustor, where there is a separate fuel split control system. Thus, pilot nozzle pressure drop and main fuel injector pressure drop are not directly related.

The design for maximum condition is: both pilot fuel and main fuel combustion are as close to stoichiometric combustion as possible. At maximum condition the overall combustion is stoichiometric to determine the maximum condition total combustion air fraction.

6.1. Combustion Air Fraction

For a combustor FAR of 0.051, combustion air is 75%. Suppose at idle condition the air flow split between pilot and main is the same as that at maximum condition. So, at idle condition the total combustion air is 75% of the liner air flow. Then at idle condition the main air flow can be determined.

For a high FAR combustor the fuel air module configuration is basically the same as the first type high pressure low emissions combustor fuel air module configuration. Both are direct mixing combustion. But for a high FAR combustor, it is not lean direct mixing combustion; rather, it is stoichiometric direct mixing combustion. For high FAR combustor, the fuel air module is shown in Fig. 21. As previously mentioned, there are three major design differences: the main fuel injection angle is less than 15°, the separation distance is less than 1*in*, and there is non-swirling air on the main swirler's outer side.

There may be applications for high FAR combustor in smaller engines with lower thrust level and lower air flow rates. The fuel/air module for such a smaller high FAR combustor is shown in Fig. 22 where, between the pilot module and main air module, there is a separation zone – with cooling air flowing through the separation zone – to allow a requisite separation distance. Note for such engine combustor application its idle LBO requirement is not strict, thus it is possible that the separation distance can be less. Fig. 22 is also suitable for lower thrust level small engine low emissions combustor.

6.2. Combustion Organization Design

For this combustor, there will be a flow divider valve. The present authors conducted a study of the flow split passing through flow divider valve and measured its pressure drop. The results show that the pressure drop through flow divider valve is nearly constant, which is called crack pressure. For combustion organization design on the main fuel and pilot fuel split, the calculation equation can be written as follows:

$$M_f = N_m * (\Delta P_{fp})^{0.5} * FN_p + N_m * N_j * (\Delta P_{fp} - \Delta P_c)^{0.5} * FN_m \quad (10)$$

Where

M_f liner fuel flow rate

N_m total module number in liner

N_j number of main fuel injectors in one module

ΔP_{fp} pilot fuel nozzle pressure drop

FN_p pilot fuel nozzle flow number

ΔP_c flow divider valve crack pressure

FN_m main fuel one injector flow number

In this equation, for different operational conditions, the pilot fuel nozzle flow number can be taken as a constant (within engineering approximation), the flow divider valve pressure drop can be taken as constant equal to crack pressure (within engineering approximation). But the main fuel injector FN for maximum condition and idle condition, cannot be assumed to be a constant. In order to define main fuel tube injector FN, it is necessary to determine its discharge coefficient C_d.

The present authors studied such plain jet tube injector. The C_d depends upon two major factors: one is tube injector length over diameter ratio, and the other is *Re* number. The main fuel tube injector in this design is relatively long, and the length was not changed significantly. Such tube injectors have a length much longer than the usual plain jet injector. Thus, the L/D is much larger than commonly used plain jet injector. This L/D ratio affects the discharge coefficient because the larger L/D will suffer more friction loss. The effect of the *Re* number is well known. The discharge coefficient will increase with *Re* number at a low *Re* number range. For a very high *Re* number range, it experiences fully turbulent flow, and C_d will not change with the *Re* number. Thus, the designer must run some specific experiments to determine the main fuel tube injector C_d (Incidentally, the discussion here on main fuel tube injector for high FAR combustor is totally suitable for main fuel tube injector in first type high pressure low emissions combustors). When there is no C_d data available, the present authors suggest using the following empirical value:

For maximum power condition,

tube injector diameter 0.02 *in*, C_d is 0.65

tube injector diameter 0.03 *in*, C_d is 0.7

For idle condition,

$$C_{d\ idle} = (Re_{idle}/Re_{max})^{0.5} * C_{d\ max} \tag{11}$$

where

Re_{idle} tube injector flow *Re* at idle condition

Re_{max} tube injector flow *Re* at maximum condition

The combustion organization design starts with a choice of a crack pressure for the flow divider valve:

a Choose a crack pressure for the flow divider valve. For example, 100 *psig*. Later, this value may be changed and will necessitate a myriad of changes in other designs.

b Design the pilot fuel-main fuel split at idle condition. For example, 70% pilot fuel flow to determine idle condition pilot fuel flow rate.

c At idle condition, design the pilot fuel air combustion equivalence ratio 1.0 to determine idle condition pilot air flow rate and to determine an idle condition pilot and main air division. Design at idle condition a pilot fuel nozzle pressure drop, say at 105 *psig* (this will be re-evaluated again when the main injector is designed), with a selected number of modules to determine the pilot fuel nozzle *FN*.

d At maximum condition assume that the air flow split between the pilot and the main is the same as at idle condition. Design at maximum condition both the pilot fuel combustion and the main fuel combustion are stoichiometric. Also, determine the pilot fuel flow rate at that maximum condition.

e Assume the pilot nozzle FN at maximum condition is the same as that at idle condition; then determine the maximum condition pilot nozzle pressure drop.

f Determine at maximum condition the main fuel tube injector pressure drop as pilot nozzle pressure drop minus flow divider valve crack pressure (assuming at the maximum condition, the flow divider valve pressure drop is the same as the crack pressure).

g At maximum condition assure the design of the main fuel combustion is stoichiometric. By this design rule determine at maximum condition main fuel flow rate. After the number of main fuel tube injectors on each module is selected, determine main fuel injector flow number, FN_m

h Select a main fuel injector diameter so the L/D ratio is defined. Design of the main fuel injection angle is the most critical step in the entire design process. It shall be taken into consideration with main fuel tube injector L/D ratio, the main air flow pattern, at maximum condition the main fuel injection pressure drop, the main fuel jet non-break length, the main fuel atomization droplet size distribution, and the number of main fuel tube injectors in each module, etc.

i By Equ. (11) the main fuel tube injector discharge coefficient at idle condition is determined. Then the idle condition main fuel tube injector flow number is determined. Now there are three parameters to be calculated: the idle condition main fuel flow rate, the idle condition main fuel injector flow number, and the idle condition main fuel injector pressure drop. These three parameters, very possible, do not agree with each other.

j Then the designer shall make some change, such as a change to the idle condition pilot fuel-main fuel split, or a change of the flow divider valve crack pressure, or a change of the maximum condition pilot fuel combustion equivalence ratio (at the same time a corresponding change to the main fuel combustion equivalence ratio. Note that at a maximum condition the pilot fuel combustion equivalence ratio should not be over 1.2 for a non-luminous flame consideration). Finally, the whole design calculation must exhibit good balance.

k Then design the main fuel tube injector co-flowing air flow rate. At the maximum condition the select main injector air-liquid ratio as two. This is a compromise between main fuel atomization at idle condition and good penetration at maximum condition.

The remainder of design work efforts, such as pilot air module design and main air module design (with non-swirling air at main module both sides), are the same as that of the first type high pressure low emissions combustor design. To summarize, the combustion organization design of next generation combustors both for high FAR combustor and for first type low emissions combustors, possess some common features:

a All combustion air comes through dome into liner and there is no combustion air from liner entry holes.

b Combustion zone aerodynamics is totally determined by fuel/air module configuration and the arrangement of modules on the dome.

c There is no axial zoning (such as primary combustion zone, an intermediate combustion zone, or a dilution zone). For type one low emissions combustors and high FAR combustors, there are a number of relatively independent but interconnected concentric pilot fuel combustion zones and main fuel combustion zones.

d They all feature direct mixing combustion. For a low emission combustor, at maximum condition, it is lean direct mixing combustion. For high FAR combustor, at maximum condition, it is stoichiometric direct mixing combustion.

e Fuel injection and co-flowing air are considered as one combination. Design the combination shall put fuel air distribution and mixing first priority, not fuel atomization first priority.

6.3. Combustion Efficiency Issue

Someone reported that high temperature rise combustors offer poor combustion efficiency issues (such as at maximum power condition the efficiency is less than 98%) because of chemical dissociation. Actually, chemical dissociation will cause combustion efficiency to decline significantly only at even much higher combustion temperature. For kerosene/air, at combustor FAR 0.051, if designed combustion is stoichiometric, combustion efficiency should be 99.7%. The explanation is as follows:

The dissociation chemical reactions in combustor are CO_2 dissociated to CO and O_2, and the water vapor is dissociated into hydrogen and oxygen, or OH. These chemical dissociation reactions will significantly reduce the heat release at temperatures higher than 2700 K (such as rocket combustion temperature), which occurs several hundred degrees higher than kerosene and air stoichiometric combustion temperatures. Based on theoretical calculation for chemical equilibrium combustion products, for kerosene and air combustion chemical dissociation may cause combustion efficiency reductions of 0.3%. This is the reason the present authors' target for high FAR combustor efficiency shall not be less than 99.7%.

This issue has some meaning for further discussion. For a gas turbine combustor, its combustion inefficiency comprises two elements: chemical inefficiency and physical inefficiency.

Chemical inefficiency is related to chemical dissociation, and it depends only on combustion temperature and pressure. If combustion pressure and temperature are defined, a combustor designer can do nothing to improve the chemical inefficiency. Physical inefficiency is related to fuel atomization (large droplets), poor fuel air mixing, not enough combustion length, fuel impingement on walls, etc. The key point is that physical inefficiency is related to combustor design and operation. Based on this discussion, there is a problem related to the method for determination of efficiency in combustor testing. According to SAE ARP 1533 the combustion efficiency is calculated by the following equation:

$$\eta = [1.00 - 4.346 * (\text{EI CO}/H_c) - (\text{EI } C_xH_y/1000)] * 100 \tag{12}$$

where

η combustion efficiency

EI CO measured CO emission index

EI C_xH_y measured UHC emission index

H_c fuel heating value

The measured EI CO and EI C_xH_y will include chemical dissociation product. Thus such combustion efficiency will not truly represent the inefficiency caused by combustor design, manufacturing and operation. In the future, the definition of efficiency determined in combustion testing shall be modified by including a computer program to on-line calculate how much EI CO and EI C_xH_y are produced by chemical dissociation. These EI CO and EI C_xH_y should be derived from the combustion test measured values, then the (physical) combustion efficiency (or inefficiency) is determined. This is the efficiency value truly representing the combustor quality.

6.4. Exit Distribution

As previously reported, a high FAR combustor exit radial profile requirement is different from that of a conventional combustor. From Van Erp and Richman (1999), the difference between temperature defined and FAR defined radial distribution profile is shown in Fig. 23.

If this truly is the situation in an engine, that is very harmful. Because of the overly rich combustion gas entering turbine meeting with turbine blade cooling air, there will be "afterburning" – totally destroy turbine blade local cooling. But this paper was twenty years ago. At that time, combustion organization was poor. Combustion was very non-uniform. According to the present authors' design, the combustion is well organized, and there will not be such severe differences. The temperature defined exit radial profile and FAR defined exit radial profile will be of the same shape. For high FAR combustor development one must always check the exit temperature profile first, then check the exit FAR radial profile. If exit temperature radial distribution is desirable, then check the exit radial FAR profile. No matter the temperature radial profile or FAR radial profile, they always are related to combustion organization. Once the combustion is well organized, these two distribution profiles will not differ significantly. Improvement of combustor design will reduce such differences. But the present author will agree that for a high FAR combustor, both temperature defined and FAR defined exit radial profile must be evaluated. Temperature defined radial distribution profile is still useful, for example, to evaluate the temperature radial

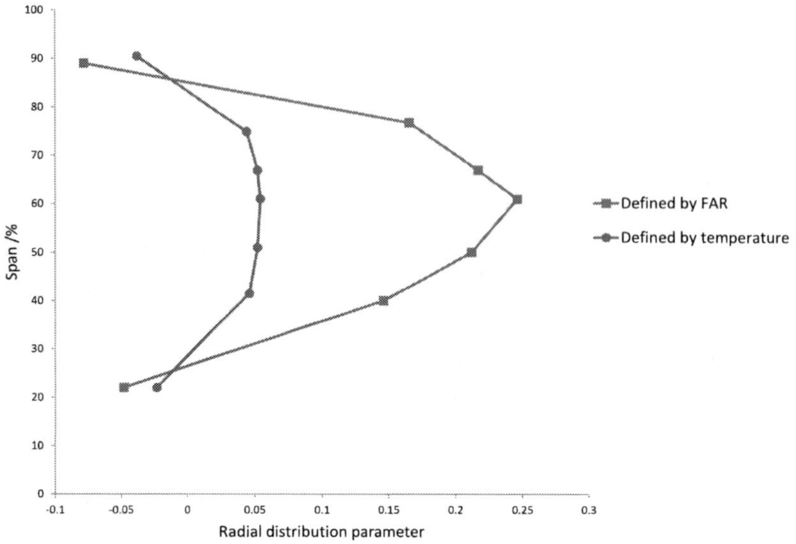

Figure 23. Difference between temperature defined and FAR defined radial distribution parameter.

profile at a lower than maximum condition first, then at the maximum power condition (probably in sector combustor testing), one can evaluate FAR radial profile just once (because it is very expensive).

6.5. Cooling Design and Calculation

High FAR combustor cooling design is very much related to combustion organization design. Because for a high FAR combustor it is not so easy to achieve a totally non-luminous flame. Thus reducing the flame luminosity is the key issue for combustion design and for cooling design, particularly with pilot fuel air combustion so it shall not be overly rich. If pilot combustion is overly rich, then the luminosity factor, which is defined by luminous flame emissivity over non-luminous flame emissivity (or gaseous phase emissivity), will be much higher than one. The present authors recommend one empirical equation to estimate the luminosity factor:

$$\frac{\varepsilon_l}{\varepsilon_g} = \phi_{pilot}^{[1+0.208*(P/10)^{0.5}]} \tag{13}$$

where

ε_t luminous flame emissivity

ε_g non-luminous flame emissivity

ϕ_{pilot} pilot fuel air combustion equivalence ratio ($\phi > 1$)

P combustion pressure, *bar* ($P > 10$ *bar*)

This equation has shown that the luminosity factor increases slightly with increase of pressure, but when the equivalence ratio is high the pressure effect is greater. Non-luminous flame emissivity is obtained by the calculation method proposed by the present authors which will be reported later. For a pressure of 200 *psia*, the comparison between the experimental luminous factor and calculated values is shown in Fig. 24. Note that this is only the effect of pilot fuel combustion richness, while pilot fuel is only one small portion of the whole combustion process. In some literature, it was suggested that the luminosity factor may be between 1.5 to 4. For the combustion organization design suggested in this book, it is not possible to have a luminosity factor of 4. Again Equ. 13 shows the importance of controlling pilot fuel combustion equivalence ratio which explains why the present design suggests that under all operational conditions pilot fuel combustion shall not be overly rich. Suggestions for high FAR combustor designs utilizing the luminosity factor 1.3 to have liner cooling calculation before any combustor testing. For a low emissions combustor the luminosity factor may held as one.

For the liner wall material, if it is possible, one may use CMC as the best choice. Hopefully, a CMC material capable of withstanding 2700°F (for short time period) will be available in the future. If this temperature cannot be reached one may consider the use of coating. The cooling hole configurations are the same as that for low emissions combustor, but laser drilling tiny holes on coated CMC liner wall remains a challenge.

6.6. Liner Cross Sectional Area

High FAR combustor usually requires very good high altitude ignition, which is a more strict requirement than that for a civil combustor. In order to meet the high altitude ignition requirement, it is necessary to design a larger liner cross sectional area. For civil combustors the design rule is that liner cross sectional

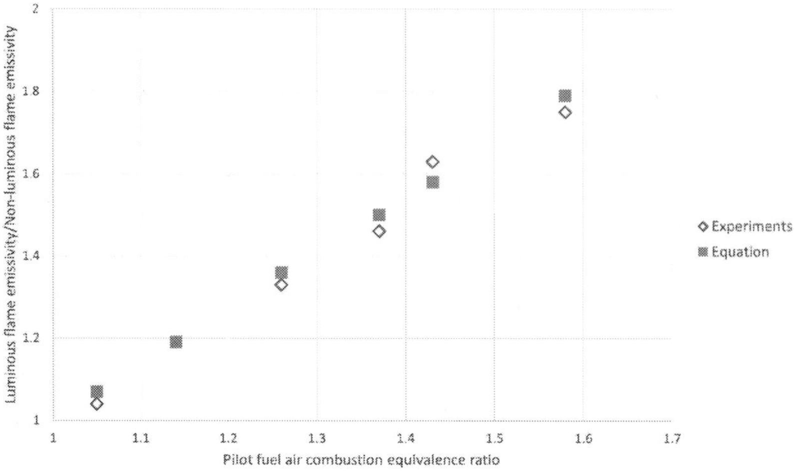

Figure 24. Luminous flame emissivity factor as a function of pilot fuel air combustion equivalence ratio ($P = 200$ *psia*)

area is equal to 12 times combustion air AC_d. For a high FAR combustor that liner cross sectional area it shall be 14 times combustion air AC_d.

6.7. NO_2 Issue

A high FAR combustor suffers a detrimental issue, and that is NO_2 exhaust. This is an environmental issue, NO_2 is toxic. When exhausting into the atmosphere, NO_2 combines with water vapor to form HONO and HNO_3. Both are volatile particulate matter compounds. But for a military combustor, it is also related to visible exhaust.

NO_2 is a brown color gas. At 50 *ppm* volume concentration it is visible. It has been observed in previous aero engine operation with the afterburner working. Thus, the design requirement is: combustor exhaust raw NO_2 concentration must be lower than 50 *ppm* (not converted to 15% oxygen concentration). In a main combustor the combustion reaction mainly generates NO, but under some conditions NO will be converted to NO_2. In any combustor, the chemical reaction of NO plus HO_2 yields NO_2 plus OH. This will be the case when high temperature combustion gases meet cold air at temperature of 1100°F. Particularly in the combustor if there is some unburnt hydrocarbons, the hydrocarbons will accelerate the NO_2 formation reaction. Thus, if the designer seeks to control NO_2 emission from high FAR combustor, three factors must be considered:

a The management of the reduction of the total NO_x levels. This proves to be very difficult. During stoichiometric combustion NO_x is at a 1000 *ppm* level. With a normal NO_2 formation rate, which is about 8% of NO_x, NO_2 is at an 80 *ppm* level.

b Avoidance of direct contact between high temperature combustion gases with cooling air.

c Attempts to reduce unburnt hydrocarbons.

Also note that when sampling gases, the sampling probe is usually water cooled. Calculation has shown that sampled gaseous flow in the sampling probe will form additional NO_2.

It is not easy to achieve a high FAR combustor design with exhaust soot particulates and NO_2 together much lower than a visible level.

6.8. Modification of Existing Combustor

There exist many aero combustor designs with a FAR not as high as 0.051, but have achieved 0.038 or lower. It is desirable to use the technology reported herein to modify those designs. The only difference is that for FAR 0.038 and lower, the combustion air will be 55.9% (or even less) and non-combustion air 44.1% (or above). There is no need to have too much cooling air because such excessive cooling air will lead to CO increases and is harmful for management of the exit profile. Only use sufficient cooling air, the rest will be dilution air.

When dilution air and cooling air together comprise 44.1% or more of inducted air, when using an air bleeding diffuser as shown in Fig. 16, the design needs to bleed more air out of the diffuser walls, and there need to be more than two bleeding slots.

Chapter 7

Next Generation Aero Combustor Development

The definition of combustor development is such that after the initial design of an aero combustor has been completed, there must be testing and then changes of the design; then retesting until the combustor operation is in agreement with all requirements. For any combustor technology development program the target is to achieve a combustor meeting the required technology target. For an engine combustor there will be ground engine test and flight bed test and finally certification. Up to now aero combustor development still remains mainly experimental work.

7.1. Technology Readiness Level (TRL)

An excellent document describing technology readiness levels, which was modified by NASA, addresses low emission combustors (NASA 2004). This document is totally applicable to military aero combustor development. It defines the technology readiness levels (TRL) from level one to level nine. This document was authored many years ago, and the present authors shall introduce that TRL document and then will suggest some modifications.

Level one is basic principle observed and reported.

Level two is technology concept and/or application formulated (candidate selected).

Level three is analytical and experimental critical function, or characteristic proof-of-concept.

Level four is component and/or bread board test in a laboratory environment.

Level five is component and/or breadboard test in a relevant environment.

Level six is system/subsystem model or true dimensional test equipment validated in a relevant environment. At this step the research program stops and moves to industry which then applies the technology to their product(s) (that is engine).

Level seven is system prototyping to be demonstrated in-flight environment capabilities.

Level eight is the actual system completion and the "flight qualified" certifications through tests and demonstrations.

Level nine is actual system "flight provedness" in operational flight.

In NASA (2004) there are explanations concerning the tests and analyses required to demonstrate TRL 1 - TRL 9. From these explanations combustor engineers acquire a greater understanding of combustor development.

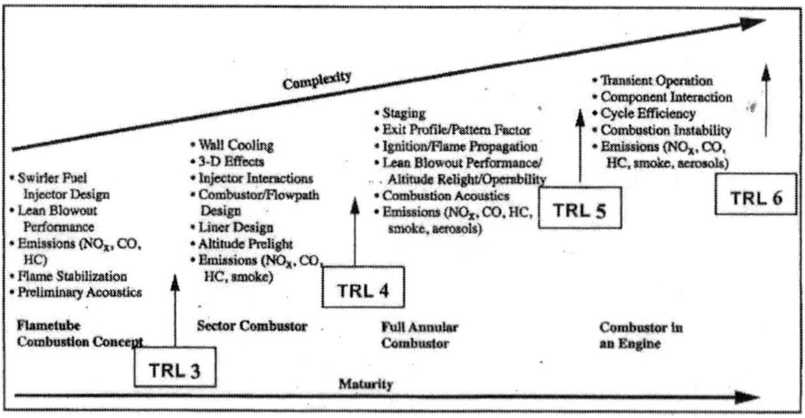

Figure 25. Combustor development process (From NASA (2004)).

Fig. 25, found in NASA (2004) provides a general overview of one of the most essential portions of the combustor development process and shows the complexity of the required testing that must increase as a technology becomes more mature. This protocol is closely related to the TRL scale. For example, flame tube testing generally corresponds to the activity required to complete TRL 3; while sector combustor testing corresponds to TRL 4; and full annular combustor testing corresponds to TRL 5, and combustor in engine testing corresponds to TRL 6.

In NASA (2004), in order to clarify each process, some specific examples from previous low emission combustor programs are listed.

> TRL 1 - This step can be as simple as a report from a basic research program that a new method to reduce emissions exists.
>
> TRL 2 - The concept is shown as it might appear in a specific application. The main objectives of TRL 2 are to eliminate concepts that clearly are not feasible and to identify other concepts that show sufficient promise to merit further development.
>
> TRL 3 - NASA usually starts working directly with industries at this stage. In the original document it is mentioned that TRL 3 tests typically are conducted in idealized "flame tubes".
>
> TRL 4 - NASA (2004) mentions that in order to fulfill TRL 4 it is necessary to start building hardware that is a reasonable simulation of real engine hardware. In a TRL 4 test its purpose is to demonstrate that the fuel injector and combustor structure can be protected with available cooling. TRL 4 tests are typically conducted in sector combustor rigs that may offer a fairly accurate simulation of the combustor cross-section. In NASA (2004) it is mentioned that the sector usually uses between one and five of the annular combustor fuel nozzles. But this is not correct. Very few aero engine combustors feature only 4 fuel nozzles. If so, such an aero combustor is definitely for a very small engine that has no need for sector combustor testing. The present authors once inspected a sector combustor design of a 12 fuel nozzles combustor by another designer. The sector had only one fuel nozzle, and it was a 30° sector. That designer stated it was a geometrical simulation. That design is exceeding bad aerodynamics. For the combustor designer it is important to gauge the aerodynamics

correctly, and not rely on geometrical simulations. Even more so, the single fuel nozzle sector combustor is a bad geometrical simulation because two side walls in 30° sector make the geometry very different. TRL 4 tests, whenever possible, are typically conducted at true engine combustor maximum condition inlet temperatures, pressures and FAR, but sector rigs are selected only to reduce the size and complexity of the required test facility.

TRL 5 - In NASA (2004) it is mentioned TRL 5 includes a range of tests including full-scale testing and full annular combustor testing typically structured to fit within an existing product engine design. For next generation aero combustor designs this testing paradigm is often feasible. Full annular tests are needed to understand combustor behavior during light-off and lean blowout, and to measure variations in temperature at the combustor exit. Full annular testing is also necessary in some cases to evaluate the effects of combustor staging on emissions. For NASA, LPP combustor development program for aircraft was terminated after TRL 5 due to the relatively high risk and catastrophic consequences of autoignition in an aircraft application. However, much of the technology demonstrated during the LPP combustor development effort was later to achieve substantial NO_x reduction in industrial engines.

TRL 6 - In NASA (2004) it is reported that tests conducted in TRL 3 through TRL 5 are conducted in test rigs at steady state or quasi-steady state conditions. To demonstrate TRL 6 capability testing is typically conducted on a slave engine. This is the first time that the combustor can be evaluated in real transient conditions (starts, engine acceleration, and deceleration) and is the first opportunity to evaluate interactions between the combustor, the compressor, and the turbine. Combustor effects on engine performance and operation at sea level can be measured directly, and durability can be evaluated by running a series of simulated mission cycles.

NASA (2004) reported that it may be necessary to make compromises in any new combustor technology to demonstrate such technology on an existing engine. In some cases there can also be mismatches between the new combustor technology and existing engine hardware. For example, the turbine cooling flows might not be optimized for the combustor exit temperature patterns of the new technology combustor, or the existing engine control might not have the capability to control all combustor functions. The present authors will discuss this issue later.

TRL 7 - In NASA (2004) it is reported that TRL 7 testing potentially can be conducted either in an altitude facility or in a flying test bed. These tests provide the first real opportunity to evaluate the combustor in the engine at air start and high altitude flight conditions. In either case, this testing is significantly more expensive than TRL 6 tests. Lean staged low emission combustor technology was evaluated through initial TRL 6 testing under the NASA program, but the potential payoff of the technology did not justify proceeding with TRL 7 testing at the conclusion of those programs. However, the technology later showed promise for military applications, and initial TRL 7 testing was conducted under a military engine development program. This clearly shows that civil combustor technology development and military combustor technology development can help each other, similarly as shown in Fig. 3 and Fig. 21, 22.

But TRL 7 testing is most likely to occur on a new engine program, where the engine can be designed to accommodate the geometric needs of the new combustor technology and the turbine section can be designed specifically for the exit temperatures produced by the new technology – and is even possible with new fuel control system. For engine combustor certification it is necessary to define the high-altitude ignition. At different altitudes, different flight Mach numbers, the combustor can be ignited by wind milling or by a starter operation, and such high-altitude ignition envelopes will be well defined. Such ignition envelopes are even more important for high FAR combustor. Certification is critical in the whole development process. In addition to emissions and other performance measurements, operational parameters must be measured. There is a combustor lifespan test, which has never be performed in the development process before. The engine runs several hundred scheduled cycles. The scheduled cycle is not equal to a real world flight cycle; rather it is a modified cycle. For each operational condition, the timing is usually shortened. This is to evaluate its long-life capability.

TRL 8 - NASA (2004) reported that TRL 8 testing is conducted to meet all applicable engine and aircraft certification requirements.

TRL 9 - NASA (2004) also reported that a significant change in design of a combustor in an existing engine can lead to an initial reduction in durability of the combustor and/or the turbine. Durability issues may not be apparent until the engines have been operated with the new component design in service and in continuing factory endurance tests for several months or years. If serious durability issues arise, the expense to solve the issue, the certification of the change, and the upgrading of the production line and the in-service fleet can exceed the total combined investment in TRL 1 through TRL 8. This is to say, combustor development never should be hurried. It must be a step-by-step deliberative process. If any previous step did not bode well, there then is no need to hastily jump to the next step.

After introducing TRL 1 to TRL 9 the present authors' intentions are to offer some suggested modification of TRL 3 to TRL 6 testing. These suggestions are offered because a long time period has elapsed since that document was published.

For combustor engineers the most important are level three to level six, as shown in Fig. 25. They are:

a Single fuel air module combustor development

b Sector combustor (usually 90°) development.

c Full annular combustor development

For technology development program, usually full annular combustor test will be the last step.

d ground engine combustor testing (combustor is integrated with engine) on an engine test bed

7.2. Single Module Tubular Combustor Development - TRL 3

For next generation aero combustor, the single fuel air module tubular combustor development is of much more importance than before. The single fuel air module tubular combustor will be the first test case in which the designer can "feel" the 100% power condition combustion. It is absolutely necessary to run the single fuel/air module tubular combustor up to or even exceeding the maximum power condition (such as FAR is a little bit higher than 100% condition). But before running up to that maximum condition there are many tests which must be performed.

A single flame tube is also useful for study of flow pattern as shown in Fig. 4, and to measure fuel/air mixture profiles by laser diagnostics both for a fundamental understanding of the combustion concept (such as direct mixing combustion concept) in the initial phases of development and to understand the behavior of full annular combustor test results in later phases of development.

The task for any single fuel air module tubular combustor development is implementing a basic combustion organization that is well conceived. In NASA (2004) it is suggested that it shall be conducted with an idealized "flame tube". This is not correct. The tubular combustor must possess a practical flame tube to run "real" test, not an idealized flame tube test. Although a flame tube is but one fuel/air module, which is a basic element of the entire combustor. For next generation combustor development it is much better to evaluate and solve the contradiction problems of lower power conditions and high power conditions during the early single tubular combustor test stage. For example, for a low emissions combustor, low NO_x is required at high power condition, but yet offers flame stabilization at any low power condition, plus offering idle LBO and good efficiency. In a single tubular combustor test one must determine whether it is possible to avoid fuel staging, or whether it is possible to use fewer stages. For a high FAR combustor the requirements are for high power conditions wherein there be no visible smoke, there is high efficiency, and reasonable liner metal temperatures; while at low power conditions there shall be good flame stabilization, and also good idle LBO and high altitude ignition. Single tubular combustor testing may evaluate transient operation. For first type low emissions combustors, testing under all transient conditions can be performed by separating the control of pilot fuel and the main fuel flow, and to evaluate what the suitable pilot fuel and main fuel combustion splits should be, as shown in Fig. 9. Also, assuming there is a flow divider valve for high FAR combustor, the pilot fuel combustion equivalence ratio at all transient conditions could be evaluated. Thus, there is no need to run a simple flame tube test at atmospheric pressure with less than 1 kg/s air flow for determining the relative stability of any new combustor concept, which is suggested in NASA (2004). Such type of very low pressure test is of no meaning. Also, the test should not be within a squared liner (which is shown in NASA (2004)). The present authors strongly suggest that for next generation combustor, a single fuel/air module (as shown in Fig. 3 or Fig. 21) with a circular flame tube and a tubular combustor should be used.

The development of a single tubular combustor for first type low emissions combustor and high FAR combustor, to large extent, will concentrate on the design of combining the main fuel tube injector and co-flowing air. The main process of air aerodynamics development is highly involved. Pilot fuel/air combustion is usually very good with no problem. Very often, idle LBO is good. Thus, development needs more effort on high power conditions. It is always necessary to monitor combustion instability levels. On single tubular combustor test, expressed by RMS, the dynamic pressure should not exceed 0.5% of the steady pressure. Later on one can compare the measured instability level with the results from sector combustor testing and full annular combustor testing. Although for next generation aero combustors, as they feature direct mixing combustion, no premixing, combustion instability usually is not a serious problem. Hopefully there may be some differences between tubular combustors and full annular combustors, but may not be significant differences. It is expected for the next generation of aero combustors, they will be free from severe combustion instability. Combustion stability is one significant advantage of LDM combustion over LPP combustion.

The final results from single module tubular combustor testing should demonstrate that all combustion performances expectations and operational requirements are met. For example, an idle LBO FAR should be at 0.006 level.

For some combustor engineers, the test data for idle LBO FAR are not so consistent. If idle LBO testing occurs ten times, the data may vary from 0.006 to 0.007. The present authors have run idle LBO tests and obtained very closely matching data, such as 0.00577, 0.00576, 0.00574. There needs to be special attention to achieve such exacting repeatability. Experience has shown the following approaches are necessary:

a Idle LBO FAR test starts from well controlled idle conditions. Inlet temperature is controlled in the range of plus minus 2°F. Pressure is controlled in the range of plus minus half psia.

b Very often the operator will simply reduce the fuel flow to lead to flameout. This is not correct. With reduced fuel flow, the combustion pressure decreases and $\Delta P/P$ increases. Thus, when at final flameout, the combustion pressure differs greatly from an idle condition pressure. That leads to different idle LBO FAR values. The present authors used combined steps such as: fuel flow reduction, increased air flow, and always maintain a constant combustion pressure.

c After FAR reaching 0.007, every control steps must be incrementally small. Every steps then in fuel flow change is only 2% of the instant flow, then the change of air flow is only 2% of the instant air flow. If the control step is large, there then is no way to define what exactly the flame-out FAR is. For example, only a one step reduction from 0.007 to 0.0058 results in flameout, and it then is uncertain whether the real flameout FAR is close to 0.007 or if it is close to 0.0058.

d Utilizing video cameras to determine flameout.

TRL 3 should solve combustion organization issue. Successful TRL 3 testing is the end of combustor preliminary design-development stage. But single fuel air module tubular combustor testing will exhibit some limitations.

a There is no diffuser, and the liner inlet flow field differs greatly from the flow field formed by the compressor exit to the flow field formed by diffuser. Only liner total pressure loss can be checked.

b Liner cooling is not the same as in full annular combustor, although the same cooling hole configuration is featured as that is in the outer liner. Liner wall temperature test data are valuable, but only for reference. If liner wall temperatures are very hot in single module tubular combustor test stage, that is an alarming signal, and that cooling must be improved (this is very rare). Since combustor durability issues are always caused by the dome temperature and liner wall temperatures and the axial or circumferential distribution, the liner wall temperature should not be a problem after that engine combustor has been certified. The designer must pay strict attention to the results of the single tubular combustor test wall temperature measurement. In a single fuel/air module tubular combustor test, circumferentially arranged thermocouples (6 or 8) are useful to indicate whether circumferentially combustion is uniform or not.

c There is no way in a single module tubular combustor test to evaluate flame propagation circumferentially during engine ignition, rather it is only possible to evaluate the ignition of one fuel air module. It is invaluable to run single module tubular combustor simulating a high altitude ignition test. Again, the high altitude ignition issue should be fully evaluated at the earliest stage. If the required simulated high altitude ignition cannot be achieved in a single tubular combustor test, then there is no need to proceed forward to the next step.

d There is no way to evaluate combustor exit temperature profiles, but the tubular combustor test may provide some qualitative information concerning pattern factors if the single tubular combustor exit temperatures are measured by multiple single point thermocouple probes.

The emissions test data from single module tubular combustor tests are usually representative of the full annular combustor. Idle LBO FAR data, 30% power heavy raining flameout data, combustion efficiency data, are representative. Derived from single module tubular combustor testing, the most important result is that the design of a fuel/air module can be finalized. But the fuel nozzle used in a single module tubular combustor test is not an engine type nozzle, rather it is a tip-on-top fuel nozzle. That will not affect the usefulness of the test result. Both fuel injectors, including pilot fuel nozzle and main fuel injector, should be tested, evaluated, modified, and the design finalized via the single module tubular combustor testing.

For single tubular combustor testing, the main fuel and pilot fuel both have separate controls. The fuel/air module and the dome are separated from liner, so it is easy to change over to a different fuel/air module for testing. For air modules, the swirlers today can be manufactured by 3D printing technology. The dome is one piece with a flange plate which would be connected to liner flange plate when running the test. Slots on dome plate allow cooling air to pass through to the cooling air channel. Downstream of the liner is an instrumentation section. The present authors designed instrumentation section is shown in Fig. 26. It is also the location for the ignitor. This ignitor location is not the real ignitor position in an engine combustor (when running simulated high altitude ignition the ignitor position is at the front side). But the ignitor is an engine ignitor. The instrumentation section is the same diameter as the liner diameter and it is water-cooled. The cooling water is exhausted axially through the water jacket outlet holes for the reduction of hot exhaust gases before reaching the exhaust control valve. There are four ports on the instrumentation section: one for the ignitor, one for dynamic pressure measurement (may use infinite line method), one for the static pressure measurement, and one for the gas sampling probe. A mixed gas sampling probe is used. The present authors designed one as shown

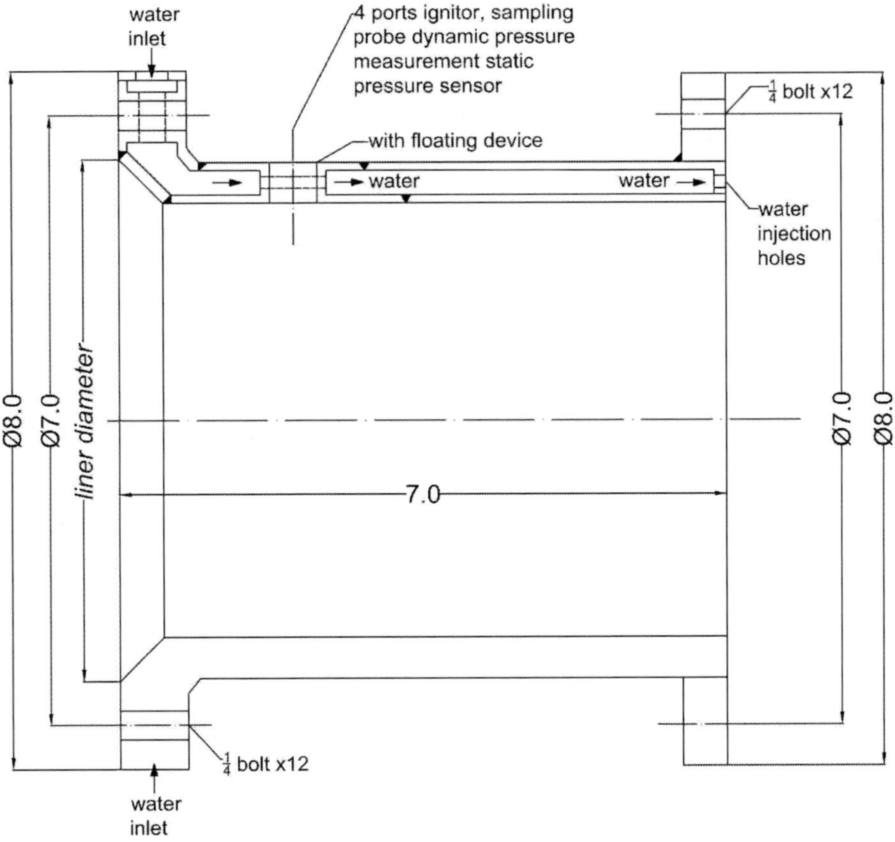

Figure 26. Water cooled instrumentation section, unit: inch.

in Fig. 27. The half inch bar material with five 0.125 *in* holes is commercially available. The central hole is used for sampled gases. On one end it is blocked. On the other end there is a water flow passage: two holes for cooling water in, and other two holes for cooling water out. Sufficient cooling for the sampling probe assures the gas sampling probe is not damaged during combustion testing – even for a high FAR combustor test.

Single module tubular combustor must run simulated high altitude ignition test. Sub-atmospheric pressure may be achieved by an ejector, whereas low temperature air is achieved by a device which makes use of high pressure air to produce lowered temperature air using a vortex tube by EXAIR Corporation.

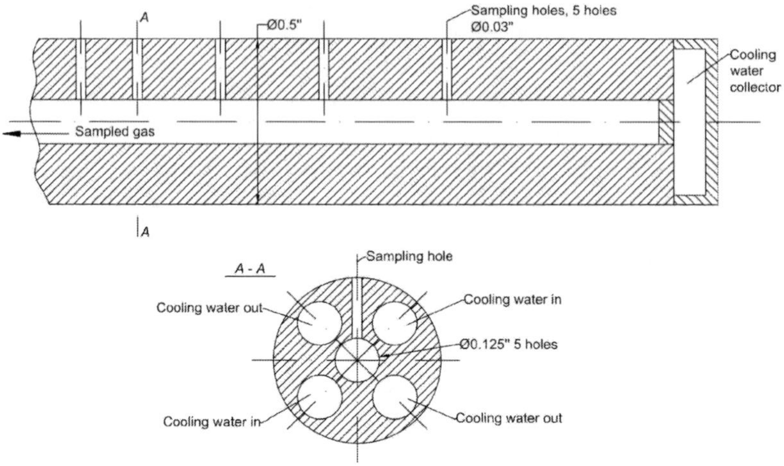

Figure 27. Mixed sampling probe, half inch 5 holes bar is commercial available.

Because there needs to run many different types of combustor test, this process usually will be time-consuming. Both pilot fuel tubing and main fuel tubing are immersed in heated air. The best way to prevent deposition forming inside the fuel tubing is to use a water cooled flexible hose. To avoid deposition, whenever a combustor test is running at a high power condition, the combustor test should not be simply shut off immediately because the dome and liner are still experiencing high metal temperatures. Suddenly shutting off causes immediate fuel nozzle deposition. The combustion test shutdown should be a deliberate process. It is necessary to maintain combustion, then gradually reduce air temperature and gradually reduce FAR, until the dome and liner wall have cooled down. Until the air temperature is reduced to about 300°F, then turn off the flame.

Past combustor development experience has shown that at the early stages of combustor development (TRL 3, TRL 4) one must attempt to detect as many problems as possible. In the early stages of development one must attempt to make use of every opportunity to evaluate the possibility of serious problems which may occur in later stages of development. At this initial stage a gross error is to enclose a single module tubular combustor test and moving forward. It shall be apparent that many problems arise concomitantly in the later stages of development.

7.3. 90° Sector Combustor Development - TRL 4

For a large engine high pressure low emissions combustor and a high FAR combustor, sector combustor development is always necessary. For small engines, but high FAR combustors, there is no need to perform sector combustor testing, and development may move from TRL 3 to TRL 5. A 90° sector combustor test is utilized to fulfill something which a single module tubular combustor test cannot achieve. That protocol is to run a maximum power condition test which the full annular combustor test facility is not able to do. Those considerations are:

a Flame propagation from one module to another module during ignition

b Liner exit temperature (or FAR) distribution profile measurements

But in sector combustor testing there is still no diffuser, and the liner cooling is still not the same as in a full annular combustor. There are two side walls. It is necessary to use water to cool these two side walls because in a full annular combustor, or in a single module tubular combustor, there is no side wall cooling air. The combustion parameters are measured using the middle 60° sector (for 3 modules combustor), or using the middle 67.5° sector (for 4 modules combustor), or using the middle 72° sector (for 5 modules combustor). A sector combustor comprises a quarter of the full annular combustor. But one full annular combustor only can be utilized to make three 90° sectors, wherein one quarter is waste. The exit temperature (or FAR) distribution measurements are achieved via a traversing gear with probes mounted on the sector outer air casing wall using a step motor with a movable sealing device. The exit temperature (or FAR) distribution measurements are not easily obtained. The present authors designed a 90 deg. sector of a rectangular shape. This is shown in Fig. 28 and also shown in NASA (2004). Experience has shown a rectangular sector combustor does not affect combustion aerodynamics, nor does such a geometry affect emissions or flame stabilization. But that design makes sector combustor manufacturing much easier and cheaper. Also, the exit distribution traversing measurement is easier.

On a sector combustor the ignitor is at an optimal position. The ignitor is installed corresponding to one fuel air module, and testing may evaluate flame propagation during the ignition process.

Sector combustors are useful for evaluating the interaction between two neighboring fuel/air modules, which in turn will determine whether the total

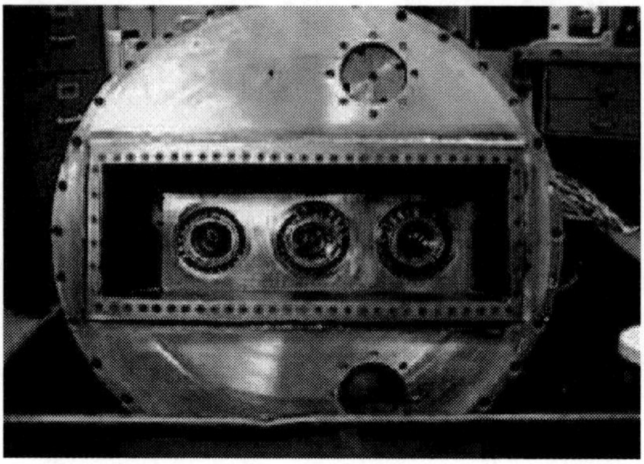

Figure 28. Sector test combustor (shown with 3 fuel injectors).

number of fuel air modules is appropriate, and also will allow the evaluation of whether there is some undesirable interference of the main air flow between two neighboring modules. For the fuel/air module shown in Fig. 3 and Fig. 21, It is critical to evaluate whether the main fuel injection is experiencing too much penetration, thus undesirable allowing main fuel overlapping in the region between two neighboring modules.

Sector combustor testing running at low power conditions in an early stages of development process provides an inexpensive means to evaluate design modifications to the combustion system. Later on, sector combustor should be tested at as high power conditions as possible. For low emissions combustor, it is to evaluate NO_x and other performances characteristics, including exit temperature distribution. For high FAR combustor, it is to evaluate high power condition efficiency, smoke, wall and dome temperatures for the cooling design, and exit temperature, plus FAR distribution. Specifically, if there is no way to obtain the exit distribution as measured in full annular combustor test, then the sector combustor test should be operated up to a maximum power condition for the exit distribution measurements. What can the designer do If the sector combustor test also cannot be operated up to a 100% power condition? The answer may be to attempt to run combustion tests at different pressures, different air temperatures and with different FAR to determine the correlation of emissions and other performances characteristics as a function of (P, T, FAR).

Sector combustor is also suitable for heavy storm weather rain ingestion flame stability testing.

7.4. Full Annular Combustor Development - TRL 5

It does not matter how high the pressure and air flow rates reach in a full annular combustor test facility. Full annular combustor testing is always necessary, unless for some technology development program only a TRL 4 level combustor test is required. For type engine full annular combustor testing, there should be an inlet air diffuser so the combustor airflow commences from the diffuser inlet. The only difference from a real-world engine combustor is that, the inlet airflow experiences a boundary layer type velocity distribution not originating from compressor exit. Compressor exit velocity circumferentially averaged offer a radial velocity profile. In a full annular combustor test it is possible to simulate such inlet radial velocity profiles if such velocity profile can be defined and really are necessary. This is the first time a designed combustor as a whole piece (with engine type fuel nozzles) appears in test rig. Only the full annular combustor offers a totally correct geometry and totally correct aerodynamics. If a 100% power condition cannot be reached, then there are two types of full annular combustor tests possible. Test results from these two tests types combined will provide information about maximum condition combustor performance and operation. Such tests are:

a Reduced power condition test. For this combustor test the uppermost condition is limited by the test facility (air pressure, air flow rate, or air heater). It is only possible to run the test up to a certain power condition (for example, say, only 65% power condition). Note that the air flow rate, fuel flow rate (then FAR), inlet air pressure, inlet air temperature, and combustor total pressure loss are all as specified by the cycle during a reduced power condition. Such test will yield the correct combustor total pressure loss, correct exit distribution, efficiency, and emissions (although at that power condition; not at the maximum condition)

b Reduced pressure test. Here the combustor test is run at the maximum power condition inlet temperature, and the maximum power condition FAR, but at a lower air pressure (also lower air flow rate). Such a test is to evaluate liner cooling, which factor is particularly important for a high FAR combustor. Because the combustion temperature is very close to, or even higher than the engine maximum conditions (if FAR is yet even higher). If FAR is the same as maximum condition FAR, because the pressure is lowered, the combustion temperature is also a bit lowered.

It then is possible to run the test at a slightly higher FAR (higher than the maximum condition FAR) to reach the maximum power condition combustion temperature. Or even higher FAR values to achieve combustion temperature higher than the maximum condition combustion temperature for evaluating liner cooling.

Very often for civil aero combustors their performance and emissions at 85% and 100% power condition can be effectively determined via the above mentioned two types of full annular combustor test results combined with data obtained from single module tubular combustor tests and sector combustor tests (with its correlation).

For a military aero combustor, if a full annular combustor cannot be run up to a 100% power condition the designer then may use the same method mentioned above to determine the performance at maximum power condition of a full annular combustor.

For next generation first type low emissions combustor full annular combustor testing, it needs to have the pilot fuel and main fuel split control system be tested before the system can be installed onto an engine. For second type low emissions combustor full annular testing, it requires to have fuel staging incorporated during the test.

For high FAR full annular combustor testing there should be incorporated a fuel system featuring a flow divider valve.

For full annular combustor testing exit distribution measurement is an issue. Exit distribution measurement usually is accomplished with traversing gear. Because such traversing measurement entail much time and are very expensive, some researchers utilize a five points thermocouples rake that continuously rotates 360° circumferentially. This is not correct because with thermocouples there is a time delay to allow readings to stabilize. Using continuous rotating method will cut off the highest temperature point and cut off the lowest temperature point. This will not affect the radial profile much, as radial profile is defined by a circumferentially averaged value, but will greatly affect the pattern factor . Such measurements will make the measured pattern factor less severe than the true value.

For traversing measurement of outlet FAR distribution, usually only one point gas sampling probe can be used (it is extremely difficult to incorporate multiple individual water cooled gas sampling probes into one unit to suit the combustor exit's limited dimensions). This entails much time and much funding. It is desirable to obtain temperature traversing measurements first, as this

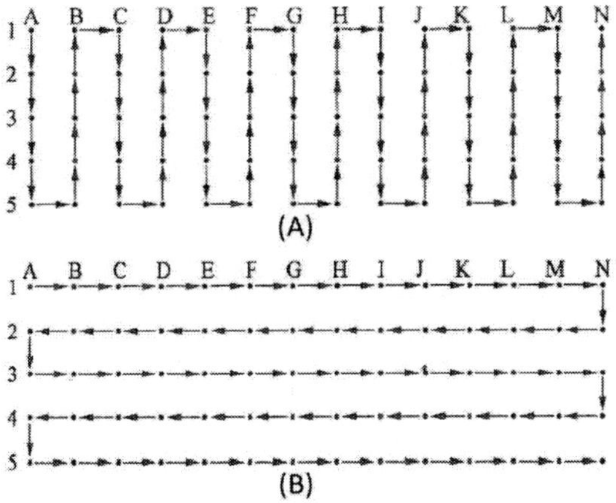

Figure 29. Change of traversing mode from (A) to (B) can save traversing measurement time.

data will provide some salient information about the exit distribution. There is also a FAR traversing issue. One way to accomplish traversing is, after taking a sample from one span position, move to another span position radially as shown in Fig. 29(A). Another way is at certain span position, traversing circumferentially, log all sample results at one span position circumferentially, then move to another span position. An example of the move is shown in Fig. 29(B). Because the gas composition differential at same span position, moving circumferentially is less than the gas composition differential changing radially. Utilizing (A) mode will require more stabilization time for a gas sampling probe to yield correct readings. Such a change of traversing mode may reduce total measurement time. Note that in order to reduce sampling time, utilizing more than one gas analyzer equipment unit is a good idea. The cost for one additional gas analyzer will be saved back in several times in such high FAR full annular combustor test.

Full annular combustor test facilities are very expensive. Running full annular combustor test is also very expensive. That is exactly the reason the present authors strongly suggest that test at single module combustor levels and at sector combustor levels as much as possible to recognize problems and solve them before any full annular combustor testing regimes begin.

7.5. Ground Engine Combustor Test - TRL 6

Ground engine combustor testing is the first opportunity the combustor is receiving true compressor exit airflow. This testing occurs in an engine test bed. The test facility can accommodate the engine thrust. Note that unless very specially installed in an engine test combustor, combustor efficiency, total pressure loss, combustor exit distribution, are not measured. Combustor emissions measurements are obtained at the engine's exit. Liner wall and air casing wall temperature measurements, if required, can be measured by thermocouples with special preparation. Usually, it is easier to evaluate combustor liner temperatures with thermal paint (it is T_{w1}). But the liner wall temperature readings obtained by thermal paint are within a range, and not any exact value; plus such approximations can be affected by the observer's judgement. Ground engine testing offers a chance to measure the fan airflow velocities (outer side of combustor casing) for future combustor cooling calculations. Also it is easy to measure combustor outer casing outer wall temperature to verify cooling calculations. Ground engine testing is the only way to evaluate combustor lifespan before certifications. Ground engine testing also is for the measurement of noise and vibration, although when engine noise is measured it is the entirety engine noise, not just combustor noise. But if a combustor is experiencing combustion instability, no matter whether of low frequency or high frequency instabilities, then that special noise frequency is definitely from the combustor. Some researchers define combustion instability as noise, which is not correct. If there is no combustion instability, the combustion noise is white noise, but combustion instability exhibited as a special frequency may be called combustion instability noise. The combustor engineer should not allow combustion instability to manifest itself when in the ground engine testing stage. For this reason, during the initial stages of development, at TRL 3 test, TRL 4 test, and TRL 5 test, combustion instability should always be monitored, and modifications should be conducted to control the instability if needed.

In NASA (2004) it is mentioned that it may be necessary to make compromises in any new combustor technology to be demonstrated on an existing engine (slave engine). In some cases, there also can be mismatches. For next generation aero combustors it will be impractical to adapt a slave engine to accept a new technology combustor for testing in an engine bed. The present authors stress one point: such mismatches may make an engine combustor test rather useless. In some cases, the mismatches are so severe that on the slave engine it is impossible to evaluate engine idle LBO. Such new but misguided combustor technology in general (such as in this book, i.e. direct mixing combustion technology, no matter lean direct mixing combustion or stoichiometric direct mixing combustion technology), is incorrect when it attempts to squeeze a new combustor design into a slave engine (such as squeezing a newly designed combustor for a pressure ratio of 70 engine into a slave engine of pressure ratio 50). Any such engine testing is mismatched. If a new engine is not available, it is necessary to use a new combustion technology to design and manufacture a new combustor for the slave engine; then they could be a good match. Then testing that slave engine with this newly designed combustor specially for that engine to see the results for the new combustion technology. This is the way, before an entirely new engine is developed, engineers can use an existing engine to evaluate new combustor technology. It is also an good opportunity to feature that slave engine with an improved combustor design. For example, for a newly designed first type high pressure low emissions combustor, the fuel system may differ greatly from a previous generation combustor design. Separate pilot fuel control and main fuel control system should be developed.

7.6. How to Run Combustor Development Test

There are many different combustor testing protocols. Here we shall only discuss the tests specifically for combustor development; not for research, not for production testing.

The following section discusses how best to perform development combustor tests correctly.

7.6.1. Combustion Test Data Must Be Repeatable

Sometimes it is found that the combustor test data are not quite repeatable. For an example, in a preliminary evaluation combustion efficiency may be 99.5%, but in a subsequent testing utilizing the same working parameters, the efficiency is only 99.2%. In order to generate repeatable data the combustor's operational condition must be stable and the instrumentation also must be stable. Before taking measurements the combustor operation should experience absolutely no change for 5 seconds. Allow the combustor to remain at that state so the operational parameters possess fewer and fewer fluctuation; then taking data.

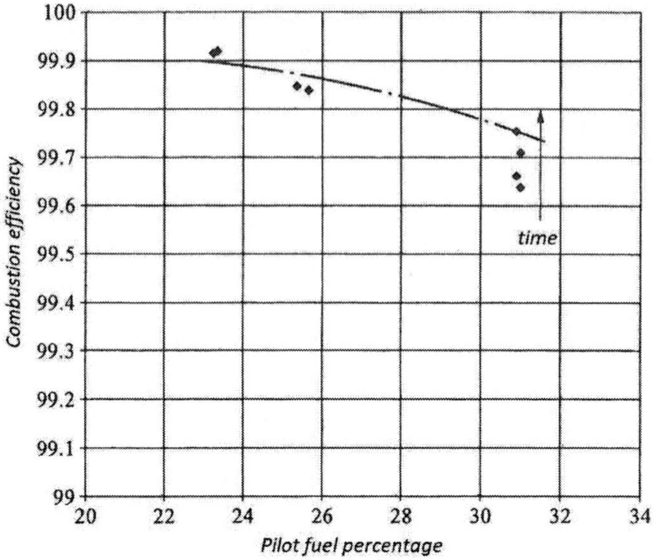

Figure 30. Because the UHC analyser is not at steady condition, measured combustion efficiency changing with time. Combustor operation does not have any change.

In addition to stable combustor operation the instruments should also be at a steady condition. NO_x instruments and UHC instruments need some time to reach a stable condition. UHC analyzers need rather long stabilization time. As shown in Fig. 30, immediately after the UHC instrument is turned on and then immediately taking the measurement, because the analyzer is not at a stable condition, the measured combustion efficiency is lower than the true value. Then within several seconds, during these few seconds, when combustor operation exhibits no change, the analyzer reading becomes stable. Measured efficiency goes up. This clearly shows that UHC analyzer must be at a steady condition, only then the measured combustion efficiency is reliable.

7.6.2. Absolutely No Air Leaking and No Fuel Leaking

For any combustor test, the test combustor or the test rig should absolutely have no air leakage. This seems rather simple, but actually it detrimentally may occur, particularly for very high pressure tests, or for sub-atmospheric pressure tests. Once there was a simulated high altitude ignition test using an ejector wherein the combustor pressure remained at 8 *psia* and there was no ignition. That dilemma seemed to be not reasonable, and the present authors had confidence that the readings should not be so poor. The present authors noticed that the combustor AC_d was declining with decreasing pressure. That is a sign of air leakage from incoming ambient air into the sub-atmospheric pressure combustor. The air flow meter reading was the same but the pressure drop across the combustor was significantly enlarged. After a thorough evaluation it was found that the technician forgot to change a gasket. Alas, he had not changed the gasket from reliable high pressure sealing to vacuum sealing. For a high pressure combustor testing air leakage causes higher AC_d falsely readings. With the same measured airflow, the pressure drop is less. For sub-atmospheric pressure testing air leakage causes AC_d lower readings. Fuel leakage causes an absolute failure in combustor testing. For tubular combustor testing or sector combustor testing, the fuel nozzle is a tip-on-top type; so there always is a connection between the fuel nozzle and the fuel line. This is the place which requires more attention. There needs to be a video camera to monitor this connection during combustor test. Even small drippage can cause combustor and test apparatus damage.

7.6.3. Take Average Readings

When running a combustor test there is, but a short timeframe for automatically obtaining any data sampling. How many samples shall be taken to obtain an averaged for one data point? The present authors performed such a study. For a stable combustor testing condition, of course all instruments must be stable. Then every 3 seconds the researcher can take a sample of FAR, and perform this for 20 times (that means 60 seconds), then derive an average of those samplings as the best practice. If one just take a reading every 3 seconds without averaging and directly make it a data point, then within 60 seconds, what obtained are 20 scattered FAR readings. Because the air flow rate readings can never be absolutely constant, the fuel flow rate readings also can never be entirely constant. These two factors together make the FAR scattered. For these twenty FAR data points, the maximum deviation from the 60 seconds averaged value is plus or minus 5%. If one uses 6 seconds to determine an averaged value (that means sample twice, and then average those readings), then there are 10 data points and the scattering is plus or minus 2%. If one takes samples 10 times and then averages those readings as a data point (so, 30 seconds averaged), then there are two data points and the scattering is only 0.2%. That is the reason why the present authors taking samples 10 times and then derive an averaged value as a data point. This process is a compromise between less scattering and shorter test times.

7.6.4. How to Judge Combustor Experiencing a Burning Out

When performing a combustor test, particularly under high power condition, inlet air temperature is high, the liner and the dome wall are hot, and then the combustion test is finalized. Is that the time to shutdown the combustor test, or shall the operator just simply turn off the test rig?

Because the combustor is in a hot state, if one is to abruptly stop the test, the fuel nozzle may form carbon deposits. The present authors once witnessed an emergency shutdown of a combustor. 30 minutes later when the test was restarted, the fuel nozzle flow number from a previous 2.5 (normal value) was reduced to 2.3. Such a situation occurs when there is a severe combustion instability, and so the test must be shutdown.

Once the present authors were present at a combustion rig to view a test in progress. The test rig screen showed that the liner AC_d was suddenly increased significantly. The rig operator turned off combustion immediately. He stated that there must be a large hole. The present authors informed him that there is no hole. The rig was opened, checked, and indeed there was no hole. Yet, the rig technician asked why it was asserted that is not a hole? The present authors explained that even a welding torch cannot burn such a big hole in a short duration of less than one second. Only because the pressure transducer is malfunctioning, was there almost no pressure drop reading. That caused a hugely increased AC_d. To make judgement about whether the combustor had burned out was not an unimportant matter. When the FAR is continuously increasing, even at any certain condition, the rig operation should force the AC_d to decrease (because the FAR is increased). But if the screen shows that AC_d is gradually going up, then it may be probable that there is a burning out.

7.6.5. How to Express Combustor Test NO_x Data?

For a low emissions LPP combustor testing several researchers prefer to express NO_x data against module FAR. This factor is expressed as liner FAR divided by the premixing module air fraction. Actually, this is not correct. Assuming two cases. One case is a liner of FAR 0.024 with a premixing of 60% module air; so the LPP module FAR is 0.04. The other case is a liner of FAR 0.032 with 80% premixing of module air; so the LPP module FAR is also 0.04. It is easily determined that the liner FAR 0.032 case will have a higher NO_x reading. Thus, it is incorrect to express NO_x data against module FAR. These two cases have vastly different temperature rises. For same module FAR, the one which experiences a lower combustion temperature rise outputs lesser NO_x.

7.6.6. Definition of Combustor Total Pressure Loss

Many combustion engineers measure exit static pressure when performing combustor testing, and only use total inlet pressure minus the measured exit static pressure as the combustor total pressure loss. This is not correct. Total inlet pressure minus exit static pressure is the pressure drop for the calculation of AC_d, and it is higher than the total pressure loss coefficient by approximately 0.3% (for a full annular combustor at high power condition). This difference will not be the same as for full annular combustor testing, nor for single module tubular combustor testing. For single tubular combustor testing, because its exit area is not contracting, the exit exhausting gas velocity is lower. In full annular combustor test the liner exit gas velocity is higher. There should be a computer program on-line to calculate under the combustor testing conditions the exit dynamic head, so to define the total pressure loss coefficient. Assuming a 0.3% difference is only an approximation, but at least it is a correct concept.

7.6.7. Change of Combustor Test Exhaust Water Flow Rate Will Affect Combustor Pressure

During combustor testing, with an increasing of FAR, there needs to be an increase in exhaust quenching water flow rates to manage exhaust gas temperatures suitable for the exhaust control valve. Often the quenching water flow rate is increased, but the combustor test operator does not make changes during the test operation. Then the combustion pressure increases with the increase of the exhaust quenching water flow rate.

7.6.8. Combustor Test to Determine the Best Main Fuel and Pilot Fuel Division

For type one high pressure low emissions combustor design (without fuel staging), one important approach is to use a separate fuel split controller for lowering emissions. That means the controller should provide the best fuel split at 30% condition, 85% condition, 100% condition, and the maximum cruise condition. But what are the best fuel splits under these conditions? That must be determined by specialized combustion testing under these power conditions.

In this case the test operation protocol should accurately maintain all other four parameters constant, including the inlet temperature, pressure, the total pressure loss coefficient, and the FAR. But only change the main fuel-pilot fuel split to define the best fuel split for low NO_X. The approach is to reduce the pilot fuel a small amount, then increase the main fuel the same quantity to maintain the total fuel flow rate constant (FAR constant), not modifying any other rig operations; so all other four parameters will remain constant with only the fuel split changed.

7.6.9. One Probe Is Not Enough for Inlet Total Pressure Measurement

For inlet total pressure measurement one probe is insufficient. For a full annular combustor test, 4 probes need to be circumferentially distributed. For the same reason the inlet air temperature also requires multiple point measurements.

7.6.10. If Any Instrument Is Out of Order, There Shall Be No Combustor Test

Not long ago, in a paper presented by one noteworthy research institute, it was reported that the UHC instrument was out of order, but that low emissions combustor test was still ongoing. The author of that paper merely assumed the UHC data were the same as in previous combustor testing regimes, which, of course, was not the same testing situation, and was using those supposed values to determine combustion efficiency. That is wrongful thinking. This is a basic scientific attitude issue. For the same reason, any original combustor test data should not be modified to suit some political needs. For combustor designer, bad data are worse than no data. Combustor testing should be run very deliberately to assure all data point measurements are totally reliable. There never should be allowed the use of "assumed" value to replace real reliable measured value to report combustor test results. If any instrument is out of order there should be no combustor test.

Chapter 8

Research on Fuel Injection and Co-Flowing Air Combination

For the design-development of next generation aero combustors, there needs to be valid research results. The following three chapters report the present authors' research on three aspects which are useful for next generation combustor design-development. Those factors are:

a Combination of fuel injection and co-flowing air

b Fuel spray evaporation

c Non-luminous flame radiation calculations

8.1. Two Types of Fuel Injection and Co-Flowing Air Combination

In next generation aero combustor there will be two types of fuel injection and co-flowing air combinations:

a Consider the combination of one pressure swirl fuel nozzle with one axial air swirler. For high pressure low emissions combustor without fuel staging and a high FAR combustor, it is a pilot fuel/air module configuration. For a type two high pressure low emissions combustor with fuel staging, it is only a basic fuel/air module configuration. Often the air liquid ratio (ALR) is quite high, such as at idle condition, its ALR is about 15.

For a low emissions combustor, at high power condition, its ALR may be higher than twenty. For a high FAR combustor, at high power condition, its ALR is also about 15. Notice that very often in other combustion research literatures that is called air assist atomization. The present authors simply prefer to define that a combination of fuel injection and co-flowing air. Because it is more than an atomization issue. Even discussing atomization under different operational conditions may be result in different atomization modes. From the combustor design point of view, the combustor designer should not merely consider how the fuel is atomized, but rather the designer should instead always consider fuel and air as a combination. Alas, fuel and air cannot be separated; so the issue is how to design that combination.

b Combination of plain jet fuel tube injector and co-axial flowing air.

8.2. Combination of Pressure Swirl Nozzle and Co-Axial Flowing Air

Firstly, lets discuss the discharge coefficient. Very often the combustor designer will assume air flow component discharge coefficient is constant. This is not true. The present authors did a study for such air swirler module with and without liquid injection, and determined that the air module AC_d varies. When a liquid injection pressure drop is low and air flow $\Delta P/P$ is high, liquid injection causes the air AC_d to lower. When the liquid injection pressure drop is high and the air $\Delta P/P$ is low, liquid injection assists higher air AC_d.

Also, fuel injector flow number (FN) is not a constant. The pressure swirl nozzle flow number will change with liquid flow rate, as shown in Fig. 31. The FN is measured in combustor tests with different inlet air temperatures and different combustor pressures. As the combustor total pressure is a constant 4%, no effect from airflow can be observed. It is obvious the nozzle flow number is increasing with the increase of flow rate.

In a separate study it is shown that without combustion, when the air $\Delta P/P$ is high and the liquid injection pressure drop is low, that air flow assists in increasing the nozzle FN. This effect is stronger when the nozzle FN is small. When air $\Delta P/P$ is low, while liquid injection pressure drop is high, airflow has no effect on nozzle FN, no matter whether nozzle FN is low or high.

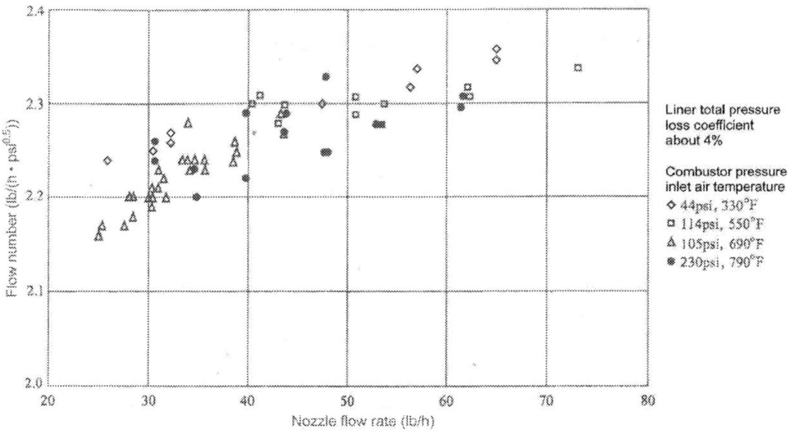

Figure 31. Combination of pressure swirl nozzle and axial swirler air module nozzle flow number measured during combustion test.

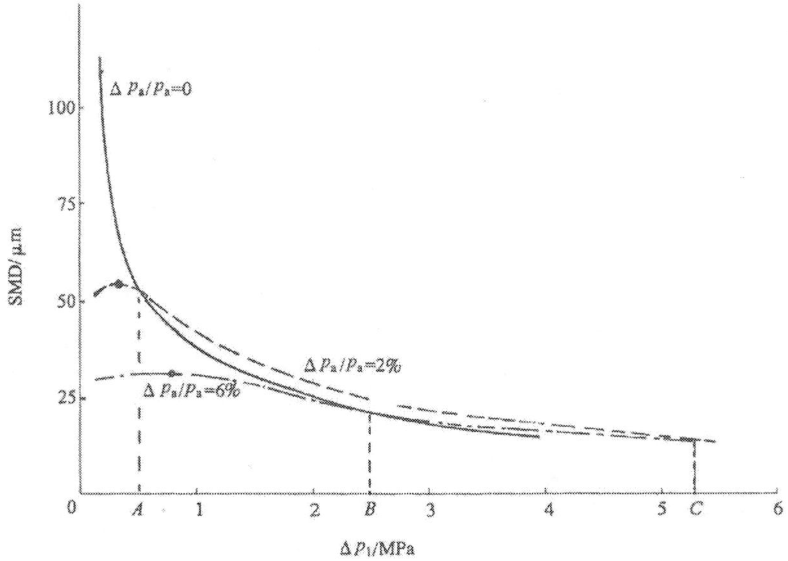

Figure 32. Atomization data expressed by SMD.

The effect of air flow on nozzle atomization is important for combustor designer and it understanding is not so simple. The present authors shall introduce

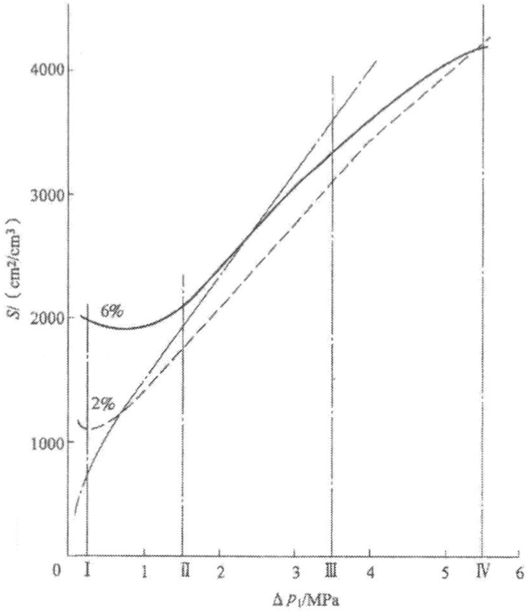

Figure 33. Atomization data expressed as liquid surface area formed by atomization for unit liquid volume S.

a parameter to represent droplet size which differs from MMD or D_{30}. It is S. This is defined as the liquid surface area increase (cm^2) by the atomization function for a unit volume of liquid (cm^3) (usually the liquid surface area before atomization can be neglected). If the unit liquid volume is cm^3 and SMD has a unit of a cm, then S is 6/SMD. For example, SMD is $50 * 10^{-6} = 0.005$ cm, $S = 6/0.005 = 1200$ cm^2. For pressure swirl nozzles and air swirler combinations, the atomization results, expressed as SMD or S values vs. different liquid flow pressure drop (ΔP_f) with three air pressure drop $\Delta Pa/Pa$ (0%, 2%, and 6%), are shown in Fig. 32 and Fig. 33. Fig. 32 is the original atomization data represented by SMD, Fig. 33, is converted to S (for clarity, the original atomization data points are deleted). In Fig. 32 and Fig. 33, when liquid pressure drop $\Delta P_l = 0.25$ MPa, with no air, or $\Delta Pa/Pa = 0\%$, $S = 700$ cm^2/cm^3; with 2% $\Delta Pa/Pa$ air, $S = 1080$ cm^2/cm^3. Within this 1080, 700 is formed by liquid pressure, liquid function is 64.8%. Thus, the function of air flow is to promote finer droplets size. But at this condition, air is not its major role, rather, air is to assist atomization, it is air assist atomization. With $\Delta Pa/Pa$ 6%, $S = 1980\ cm^2/cm^3$, at this condition, the liquid pressure function is only 35.4% and air atomization is then playing the major role; thus this is air blast atomization. At $\Delta P_l = 1.5\ MPa$ with no air, $S = 1900\ cm^2/cm^3$, with $\Delta Pa/Pa = 6\%$, $S = 2100\ cm^2/cm^3$, air only assists to increase 200 cm^2/cm^3; thus it is air assist atomization. Now check the $\Delta Pa/Pa = 2\%$ situation. Without air it is 1900 cm^2/cm^3. With 2% $\Delta Pa/Pa$, S is even less, reduced from 1900 cm^2/cm^3 to 1740 cm^2/cm^3. That means air is not assisting atomization; rather it is retarding atomization. This situation we shall call "air retard atomization". This technical term, air retard atomization, is now appearing for the first time in atomization literature.

Evaluate $\Delta P_l = 3.5\ MPa$. When there is no air, $S = 3600\ cm^2/cm^3$, with 6% $\Delta Pa/Pa$, $S = 3350\ cm^2/cm^3$, then air retard atomization is occurring here. Evaluate the $\Delta Pa/Pa = 2\%$ situation, $S = 3100\ cm^2/cm^3$, where air creates a more negative effect.

Evaluate $\Delta P_l = 5.5\ Mpa$, no matter whether $\Delta Pa/Pa$ is 2% or 6%, Both factors create air retard atomization, but at this condition 6% $\Delta Pa/Pa$ will create a more negative effect on atomization than 2% $\Delta Pa/Pa$.

For a different pressure swirl nozzle with a different FN and a different air swirler, the change of atomization mode will occur under different conditions, but there will still be three atomization modes:

a Air blast atomization mode

b Air assist atomization mode

c Air retard atomization mode

Many years ago, A. H. Lefebvre stated that air assist atomizers are characterized by their use of a relatively small quantity of very high velocity air, while air blast atomizers are characterized by their use of a large amount of relatively low velocity air. At that time the present author, Jushan Chin, raised the question whether there is any good reason why a small amount of high velocity air should be assisting atomization while a large amount of low velocity air is called air blast atomization. Lefebvre said: 'Then you give that a good definition.' From the previous mentioned atomization study, the present authors proposes a definition.

With pressure swirl atomizers and co-flowing air combination, if air provides the major positive atomization function, then that is air blast atomization mode. If air provides a minor but positive function, then that is air assist atomization mode. If air is providing a negative function, then that is retard atomization mode. For a pressure swirl atomizer and co-flowing air combination, we shall not call the combination an air blast atomizer, nor an air assist atomizer because the same combination, under different conditions, may experience different atomization modes.

This study has useful meaning:

a The combustor designer should consider fuel nozzles and air swirler modules in combination, not as merely a fuel injector. Because air/liquid ratios now are always very high, and those components have strong interaction; fuel and air should be considered as one component of combustor operation.

b The combustor designer should not merely pay major attention to atomization (droplet size). Droplet size is important at low liquid injection pressure conditions. Under other conditions, particularly at high power conditions, injection pressure drop is rather high and droplet size is not the major issue. Fuel/air mixing is more important.

c Air retard atomization is not a terrible situation. Very fine droplet size is not always good for combustors. Air retard atomization often offers a suitable droplet size distribution for fuel dispersion and fuel/air mixing. The combustor designer should recognize that for pressure swirl fuel nozzle and co-flowing air combinations the air is not only atomization air, but it is also dispersion air, mixing air, and finally it is combustion air. This is a change in fundamental design conceptualization. Thus, this book is not merely an academic research study.

8.3. Combination of Plain Jet Tube Injectors and Co-Axial Flowing Air

When combining plain jet tube injectors and co-axial flowing air for main fuel-air combustion, the air/liquid ratios are much less than the pilot fuel air combination. For main fuel tube injectors and co-axial flowing air the resulting ALR is often at the order of 2 - 3. The flowing air will not affect injector flow numbers. Injector flow number (FN) will be significantly influenced by liquid flow rate, as shown in Fig. 34. These data points are measured in a combustor test. As reported previously, for a high FAR combustor at idle condition, the main fuel pressure drop is very low; thus its flow discharge coefficient is low. The effect of airflow on the liquid discharge coefficient at high power condition is weak. The effect of the liquid flow on the airflow discharge coefficient is also weak.

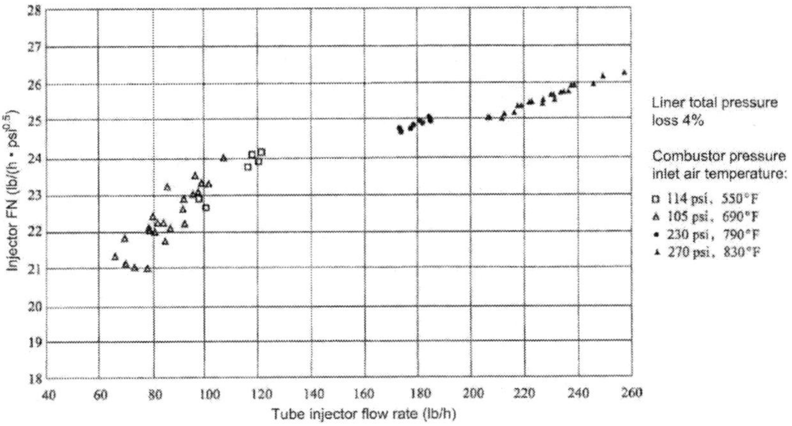

Figure 34. Plain jet tube injector and co-flowing air combination, injector *FN* changing with injector flow rate.

For such tube injectors co-flowing air combinations, most concerning is fuel spray penetration and dispersal. At high liquid injection pressure drops and high jet velocities, there is a length, before which the liquid jet will not break up. This breakup length is important to consider by combustor designer.

Note that the tube injectors and co-axial flowing air combinations are very different from previously researched plain jet air blast atomization, and differ from liquid rocket engine (liquid oxygen-gaseous hydrogen) co-axial atomization. Most importantly, present designs feature very high length-to-diameter ratios and the liquid injection pressure drop can be rather high. Many classical researches on jet disintegration are for pure liquid jet and without co-axial air flow. In rocket propellant injection there will not be so high a length-diameter ratio (L/D). But from previous research it understood that the breakup length will increase with jet velocity (injector pressure drop) and an increase in tubing diameter. Since the liquid is always aviation kerosene, its liquid property does not affect the breakup length. Higher air/liquid ratios will reduce the breakup length. In conclusion, the combustor designer needs to perform some specific studies to understand such plain jet tube injectors and co-flowing air combinations. How long the breakup length is and its related penetration; all of which factors are necessary for the designer to determine the one most important design choice, and that is the main fuel injection angle relative to the module axis. The present authors reported a value 15°, or less than 15°, as only an example. That angle should be designed after some specific study.

For main fuel plain jet tube injectors with co-flowing air combinations, there are also three atomization modes:

a Air blast atomization mode

b Air assist atomization mode

c Air retard atomization mode

As air/liquid ratios are low, air retard atomization will appear at relatively lower liquid pressure drops than that in the pressure swirl nozzle and air swirler combination.

For such tube injectors, if the liquid velocity is equal to the air velocity, the droplet size is largest. Suppose liquid velocity is 50 m/sec, if air velocity is 40 m/sec, or 60 m/sec, both have a relative velocity of 10 m/sec and both will provide a finer droplet size than that with an air velocity of 50 m/sec. But the case with an air velocity of 60 m/sec will provide even finer droplet size than the case with an air velocity of 40 m/sec. This illustrates that for atomization relative velocity is one factor which affects droplet size, while air has its own effect on droplet size.

The present authors performed an atomization research on plain jet injectors with co-axial flowing air under a high ambient pressure. Using a supersonic nozzle within a high pressure chamber similar to a rocket combustor, two side walls were equipped with optical windows, the chamber pressure varied from 1 *bar* to 16 *bar*, and droplet size was measured (Yang and Chin 1990). Water injection was via plain jet nozzle with coaxial airflow. Injection velocity was 35 m/sec. Coaxial airflow velocity was varied, i.e. 60 m/sec, 80 m/sec, 100 m/sec, 120 m/sec. Droplet size SMD decreases with increases of chamber pressure. With higher relative velocities the negative exponents of SMD with pressure were even more as shown in Table 2.

Table 2. Correlation between SMD vs. air velocity and relative velocity

Air velocity	Relative velocity	SMD
m/sec	m/sec	pressure index
60	25	$P^{-0.38}$
80	45	$P^{-0.42}$
100	65	$P^{-0.49}$
120	85	$P^{-0.52}$

The atomization research with high chamber pressures is very meaningful for next generation aero combustor development. For pressure swirl/fuel nozzle and axial swirler/air combinations, under high combustion pressures the atomization is very fine. With close contact between swirling air and fine spray, fuel evaporation is very fast and fuel/air mixing is good. That is the reason such a combination will provide low emissions at high power conditions. But for plain jet tube injectors with co-axial flowing air combinations, at high power condition very fine droplet size will not be helpful for spray penetration. That is the reason for main fuel injection design, at high power condition, managing fuel spray penetration, dispersion, and fuel air mixing is more important than atomization. When the injection pressure drop is high, the injector is a tube with diameter 0.02 *in* to 0.03 *in* and high L/D ratio, it retards the atomization. Note that the main fuel penetration shall be optimized, while insufficient penetration will cause local fuel rich combustion, while also too much penetration, or over penetration, will cause two neighboring main fuel injection cycles to experience fuel overlapping; so also causing local fuel rich combustion.

Finally there is a design choice concerning the relative positioning of liquid injector exit orifice and air flow orifice, and whether a liquid injector orifice protrudes from an air orifice, or an air orifice protrudes outwardly from a liquid orifice. In rocket combustion, such design features will affect combustion instability. For an aero combustor there is no severe combustion instability. Thus, the liquid tube injector orifice is flush with the air flow orifice.

Chapter 9

Fuel Spray Evaporation Research

The present authors have engaged in extensive fuel spray evaporation research. But before reporting on these research results, let us now discuss some basic concepts.

9.1. Some Basic Concepts

a In most combustion literature about fuel evaporation there is a statement that the evaporation of droplets in a spray involves a simultaneous heat and mass transfer process in which the heat of evaporation is transferred to the droplet surface by conduction and convention from the surrounding hot gas, and that vapor is transferred by convection and diffusion back into the gas stream.

There once was a Canadian student who raised a question. He related that in arctic Canada the winter ambient temperatures are very low, and outside air temperatures usually are much lower than fuel temperatures. Would there be any fuel evaporation? Before combustion, if there is no fuel evaporation, how possibly could a combustor be ignited?

This is a basic concept. If there is no hot air, and if the ambient air temperature is lower than fuel temperature, will there still be any fuel evaporation?

Because there is no heat transfer from such cold ambient air, where does the latent heat for evaporation come from? The answer to this question is yes! Even if the ambient temperature is lower than the fuel temperature there will still be fuel evaporation. The latent heat for evaporation comes from fuel enthalpy; that is to say that during fuel evaporation the fuel droplet temperature decreases while fuel enthalpy is also decreasing.

b In some literature it is mentioned that during droplet evaporation the droplet diameter continuously decreases.

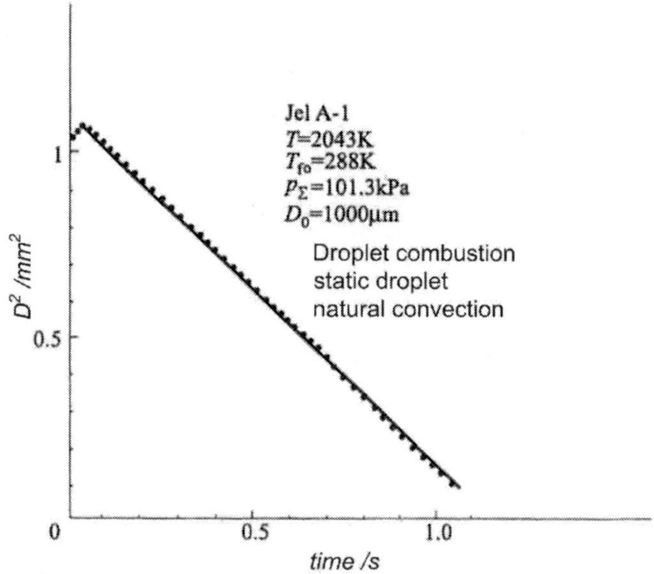

Figure 35. The present authors calculated single droplet evaporation history compared with experimental results in Godsave (1953).

It is untrue. Fig. 35 shows the present authors calculated single droplet evaporation history compared with Godsave experimental data (Godsave 1953). The dotted points are camera recorded with some conversion of volume to define a diameter. It is clear that at beginning of evaporation the droplet temperature increases to create a volume expansion more than the volume decrease by evaporation; also the diameter is enlarged, and after a certain time (droplet heat up period) the droplet diameter squared decreases along a straight line.

c Many years ago there was a researcher who conducted some measurements of droplet size Mass Medium Diameter (MMD) of a spray along a downstream distance. His results showed that MMD was increasing with downstream distance as the spray underwent more and more evaporation. He mentioned that his observation may be experimental error because during spray evaporation all droplet diameters decrease. At that time the present author, Jushan Chin, discussed this with A. H. Lefebvre. We all agreed that this was a misinterpretation of observed data and was not due to experimental error. Rather, this phenomenon is due to the fact that smaller droplets dissipate faster than larger droplets. Then, our research was published in Chin, Durrett, and Lefebvre (1984). The major conclusion is that as evaporation proceeds, the mass median diameter (MMD) and droplets size distribution parameter both increase in the evaporation process for any spray pattern with an R-R distribution parameter less than 4. This effect is especially pronounced for sprays possessing a low droplet size distribution parameter (such as the parameter of two or less than two). Only a spray with R-R distribution and a droplet size distribution parameter higher than 4 will MMD decrease via the evaporation process. From a combustion efficiency viewpoint, it is beneficial to achieve a higher droplet size distribution parameter (means more uniform spray), though such a droplet size distribution parameter may be undesirable from an ignition standpoint.

For next generation aero combustor, combustion efficiency usually is not a big problem – even at idle conditions. But for high altitude ignition, particularly for high FAR combustors, hopefully the pressure swirl nozzles and swirling air combinations will provide a spray with low droplet size distribution parameters to assist ignition.

For next generation aero combustors, it is essential to correctly predict fuel vapor/air distribution. This is very much related to the calculation of fuel spray evaporation.

9.2. Single Component Fuel Evaporation Calculation

Most aviation fuels are multi-components fuel, such as JP-4, JP-5, JP-7 JP-8, JP-8+ 100, JP-TS, Jet A, Jet A-1. They are a mixture of many hydrocarbon compounds (plus additives). There is only one single component jet fuel, and that is JP-10. Other hydrocarbon fuels, such as n-heptane, are single component fuels, but they are not aviation fuels. For multi-components fuels there is a distillation curve. It shows that at different temperatures the individual components will distillate from a liquid phase to a gaseous phase. Light components (with lower molecular weights) will vaporize (distillate) first, then heavier component will vaporize (distillate) later.

Fuel droplet evaporation research started from single component fuel evaporation. About 40 years ago the present author, Jushan Chin, and A.H. Lefebvre proposed a single component fuel evaporation calculation method (Chin and Lefebvre 1983a). In that paper we mentioned our single component fuel evaporation calculation, which relates to two things:

a Single component fuel such as n-heptane, or JP-10 to calculate droplet evaporation.

b The calculation method is used for multi-components fuels; that is, using a single component fuel droplet evaporation calculation method to approximately calculate multi-components fuel evaporation. That method utilize a 50% distillation point property to represent multi-components fuel. This is simplified multi-components fuel evaporation calculation method. Here is is to consider any multi-components fuels as a single component fuel.

Actually, our fuel evaporation research was very much promoted by the development of lean premixed pre-vaporized combustors. About 40 years ago the single component fuel evaporation calculation methods were reported in Chin and Lefebvre (1983b) and Chin and Lefebvre (1983a) for steady state evaporation and droplet heat-up period evaporation calculations. When fuel spray evaporation occurs in a gas turbine combustor, usually the whole spray is divided into at least 10 droplet size groups (though it is better to consider 20 drop size groups). In each size group the researcher should choose an average diameter to represent the fuel mass for this size group. For each size group of droplets simply calculate the representative droplet evaporation history and movement trajectory. Thus, the whole spray evaporation problem becomes one of a droplet evaporation calculation problem. For CFD it matters not what the turbulence model is; there must be a fuel spray (droplet) evaporation model. This evaporation model must include droplet heat-up period and steady state evaporation, and include convective effect and a droplet drag coefficient equation. The evaporation calculation must use correct thermal physical properties, including liquid phase properties, gaseous phase properties, latent heat, etc. The detail physical property calculation is not provided herein. Recommendation for readers is to use those salient details reported in Chin and Lefebvre (1983b) and Chin and Lefebvre (1983a).

Chin and Lefebvre (1983a) and Chin and Lefebvre (1983b) combined together are used by many researchers and combustion engineers. One of them is cited in Dodge and Moses (1984). In Dodge and Moses (1984), the authors mentioned: "Much of the computer model was based on work done by Professors J.S. Chin and A.H. Lefebvre at Purdue." The present authors of this book will not report the details of the calculation method herein.

Chin and Lefebvre (1983b) and Chin and Lefebvre (1983a) were research papers published 40 years ago. Up to now those methods described therein were correct. But at that time there were not much reliable fuel vapor pressure experimental data; thus, the vapor pressure correlation equation then was as follows:

$$P_{fs} = \exp\left(a - \frac{b}{T_s - 43}\right) \tag{14}$$

where

P_{fs} vapor pressure, kpa

T_s droplet surface temperature, K

a, b constants with different fuels

Gauthier, Bardon, and Rao (1991) presented multiple fuel vapor pressure and molecular weight data for three fuels. The present authors have checked the 50% distillation point vapor pressure determinations from Gauthier, Bardon, and Rao (1991) with Equ. 14, and there is some disagreement. That means, using Equ. 14 to calculate vapor pressure, a 50% distillation point vapor pressure is not entirely correct. The present authors have made use of vapor pressures of three fuels from Gauthier, Bardon, and Rao (1991) to correlate a new equation for simplified multi-components fuel evaporation calculations. That equation is as follows:

$$P_{fs} = \exp\left(A - \frac{B}{T_s}\right) \tag{15}$$

where

P_{fs} vapor pressure, kpa

T_s droplet surface temperature, K

A, B constants

For Jet A-1, JP–4, and DF–2 fuels the constants are listed in Table 3.

Table 3. Constant A and B for Jet A-1, JP–4, and DF–2 fuels

Fuel	A	B
Jet A-1	15	5193.841
JP-4	14.75	4250.127
DF-2	15.15	5775.144

Using Equ. 15 to calculate vapor pressures at different droplet surface temperatures agree with the 50% distillation point vapor pressure from Gauthier, Bardon, and Rao (1991) very well. For these three fuels, when using a simplified calculation method to compute their evaporation coefficients, the physical properties needed are listed in Table 4.

Table 4. Physical properties for Jet A-1, JP–4, and DF–2 fuels

Fuel	Jet A - 1	JP - 4	DF - 2
T_{bn}, K	497	420	536.4
T_{cr}, K	687.9	612	725.9
P_{cr}, kPa	2213.5	3654	2089
L_{Tbn}, kj/kg	262	292	254
Molecular weight	173.5	125	198
Density (288 K), kg/m^3	820	771	846

where

T_{bn} normal boiling point

T_{cr} critical temperature

P_{cr} critical pressure

L_{Tbn} latent heat at T_{bn}

9.3. Multiple Component Fuel Evaporation Calculations

Aviation kerosene-based fuels are multi-component fuels. Their evaporation calculation should use the multi-component fuel evaporation calculation method. Many years ago several scientists tried using one component plus another component (such as A plus B), solving partial differential equations to calculate two components fuel evaporation. But that really is not the means to solve aviation fuel multiple components evaporation problem. Because for aviation fuel, to really represent the physical properties, there must be at least 33 components. Thus, there must be a realistic engineering method to solve multi-component fuel evaporation calculation problems.

Fuel evaporation research first started from single component fuel calculation, then developed up to multi-component fuel calculation. For multi-component fuel evaporation, the understanding is that when a droplet begins to evaporate, at the droplet surface the light component (of low molecular weight) will vaporize first. After the light components have vaporized from the surface, the light component concentration in the central portion of the droplet will be higher than the surface layer. How the higher concentration light components in the central portion will diffuse towards surface layer create the potential for three situations:

a The light vaporized component has zero liquid phase mass diffusivity; that is, nothing from central portion will diffuse to the surface layer. The surface layer will totally vaporize every component until all components finish evaporating; then will commence the second layer evaporation, one layer by one layer, like an onion. This type of evaporation is called "onion" mode. This is one extreme situation. This phenomena is also called zero liquid phase mass diffusivity mode.

b After a surface layer light component have vaporized the central portion light component will instantaneously diffuse the component to the surface layer to keep that component concentration totally uniform within the entire droplet. This is another extreme situation and is called infinite mass diffusivity mode. Both zero mass diffusivity mode and infinite mass diffusivity mode are thermodynamically equilibrium conditions.

c The real situation is actually between the above two extremes, and it is defined as finite mass diffusivity mode.

Fuel droplet liquid conductivity may be assumed to be of an infinite thermal conductivity or of a finite thermal conductivity. If it is assumed to be infinite thermal conductivity the entire droplets temperature is always uniform. The present authors studied this issue. The conclusion is that no matter for single component fuel evaporation or multi-components fuel evaporation, if assuming finite thermal conductivity, the calculation requires to solve an un-steady thermal conduction equation. This is much too complicated. For aero combustor design and development, that really is not necessary and is not realistic. Thus an infinite thermal conductivity assumption should be used. There is another reason to do so. One research has shown that inside a droplet, there is an internal liquid recirculation, although it is for large droplet sizes such as $200 * 10^{-6}\ m$.

For multi-components fuel evaporation, the most important information is from Gauthier, Bardon, and Rao (1991). The authors of Gauthier, Bardon, and Rao (1991) converted the distillation curve into a vapor pressure equation for three multi-component fuels: Jet A-1, JP-4, and DF-2. The equation is as follows:

$$P_{fs} = f(T_s, \text{EVAP}) \tag{16}$$

Also the vaporized fuel (vapor) molecular weight equation:

$$M_{fv} = f_{fv}(\text{EVAP}) \tag{17}$$

And the molecular weight of the liquid fuel remained in droplet equation:

$$M_{fav} = f_{fav}(\text{EVAP}) \tag{18}$$

where EVAP is the droplet vaporized fraction.

These equations have been derived from their original equations with simplified form.

Here the most important information is that for multi-components fuel evaporation, its vapor pressure and molecular weight (both vaporized and liquid phase remained) are not merely dependent on droplet temperatures, but they are also dependent upon vaporized fraction.

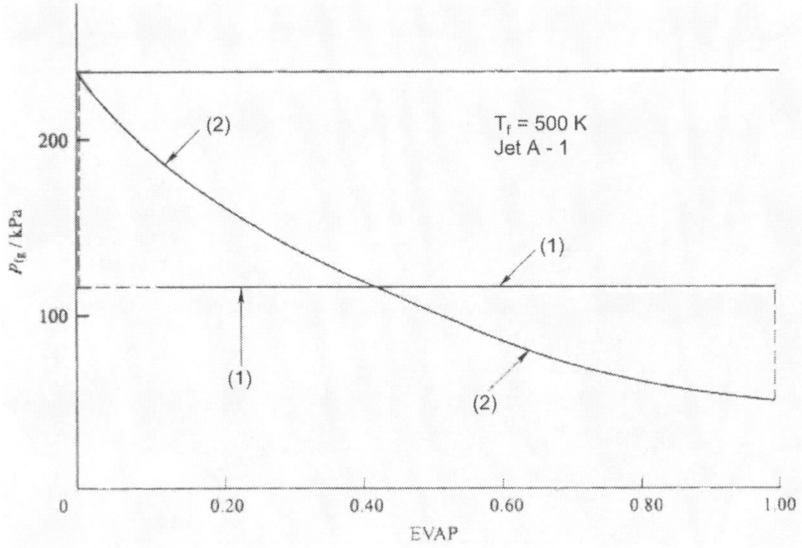

Figure 36. Jet A-1 fuel, vapor pressure calculation, at 500 *K* comparison between single component calculation and multi-component calculation with infinite mass diffusivity: (1) Single component calculation, (2) Multi-component calculation with infinite mass diffusivity.

Fig. 36 shows the comparison of vapor pressure for Jet A-1, with a surface temperature of 500 *K*, calculated by a single component evaporation method and by multi-component evaporation with an infinite mass diffusivity assumption.

Here the most important difference is that at a constant droplet temperature (500 *K*), by single component calculation method the vapor pressure is constant, and does not vary with evaporation history; while by multi-component calculation method the vapor pressure varies with vaporized fraction. Note that via a multi-component calculation the vapor pressure initially is higher than that calculated value derived from a single component calculation, whileas subsequently the vapor pressure calculated by a multi-component calculation is less than that determined by a single component calculation. Also, with multi-component calculation, during the evaporation process, the vapor molecular weight is changing with the vaporized fraction, but the molecular weight is not from only one type of molecule; rather, the molecular weight is an average of a mixture of many component molecules.

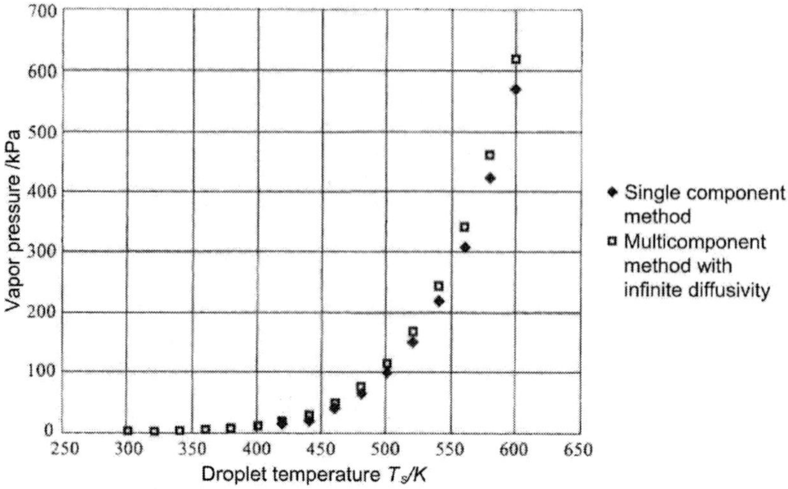

Figure 37. Jet A-1 fuel comparison of calculated vapor pressure between the single component calculation method and multi-component method with infinite diffusivity.

The present authors have conducted fuel evaporation research for many years. Some of our research results offer a practical solution for aero combustor usage, while some of our results are only for study purposes. And some of our calculations are not realistic for engineering use. The conclusions are listed as follows:

a The onion mode (zero diffusivity) multi-component evaporation calculation is very close to the simplified single component evaporation method (using 50% distillation point as being representative), as shown in Fig. 37, the maximum difference is about 5%.

b The present author has tried to use finite diffusivity to calculate multi-component fuel evaporation, and that calculation process is very complicated. The conclusion is for aero combustor engineering that calculation method should not be utilized.

c The present author has tried using finite liquid thermal conductivity to calculate single or multi-component fuel evaporation, and that too is unrealistic. The conclusion is to use liquid infinite thermal conductivity assumption.

d If only to calculate the droplet total lifetime, then using a simplified single component calculation method or a multi-component method, the difference is insignificant.

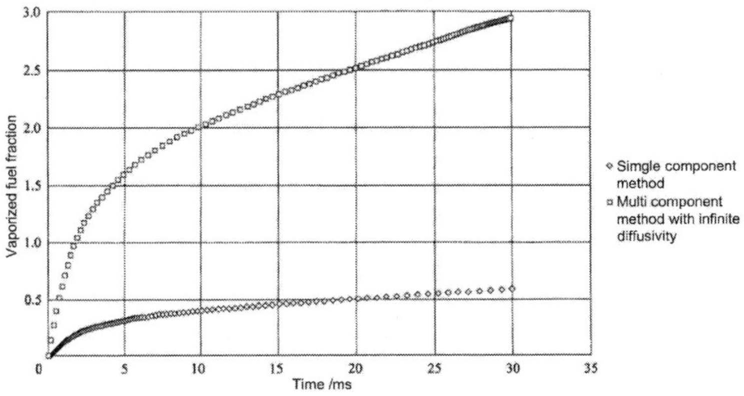

Figure 38. Jet A-1 fuel comparison of calculated vaporized fuel fraction for cold air ignition between single component method and multi-component method with infinite diffusivity (Vertical cross stream injection, air temperature $T_a = 255K$, initial dropsize of $50 \cdot 10^{-6}$ m, initial droplet velocity of 35 m/s, air velocity of 15 m/s, air pressure of 10 $psia$).

e For low temperature ignition purpose utilizing a single component calculation method will lead to significant errors. As shown in Fig. 38, the single component method calculated vaporized fuel fraction is much lower than the vaporized fuel fraction calculated by a multi-component method with infinite diffusivity. That is the reason via a single component calculation method there will be no ignition, but in the real world situation there indeed is ignition. Conclusion: for low ambient temperature ignition, multi-component evaporation calculation method with infinite diffusivity should be used.

f For combustor CFD, the designer needs to know the fuel (vapor) air distribution. Although the total droplet evaporation time is not much different

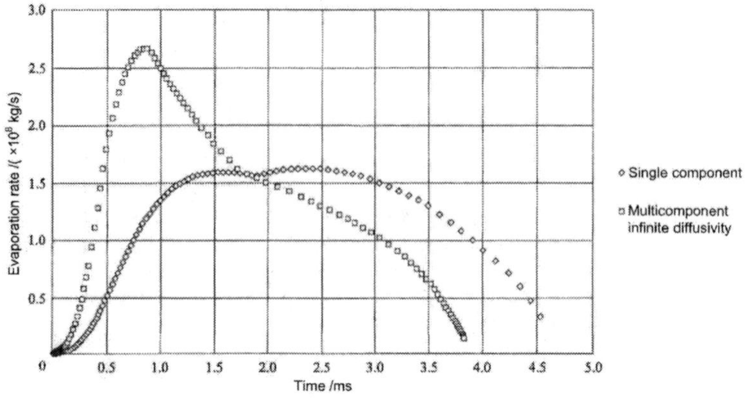

Figure 39. Jet A-1 fuel, droplet evaporation rate changing with time. Comparison between calculation by single component method and multi-component method with infinite diffusivity (droplet initial diameter of $50{\cdot}10^{-6}$ *m*, droplet initial velocity of 23.7 *m*/*s*, vertical injection. Cross stream air velocity of 50 *m*/*s*, air pressure of 300 *psia*, air temperature of 800 *K*).

for the two calculation methods, the evaporation rates for the two calculation methods is different. The multi-component method will predict higher evaporation rates initially, while the single component method will predict higher evaporation rates at the later portion of droplet lifetime as shown in Fig. 39. Thus, for CFD calculation of fuel (vapor) air distribution, utilizing these two methods will obtain rather different results.
Conclusion: for CFD to predict fuel (vapor) air distribution, the multi-component method with infinite diffusivity should be used.

g In Chin (1994b), Chin (1994a), Chin (1995) and Chin (1995), the present author, Jushan Chin, considered fuel evaporation along with droplet secondary breakup; then the factors considered were droplet critical condition evaporation, possible droplets interactions, and the effect of non-zero vapor concentrations in the surroundings. These studies were academic exercises, and are not appropriate for engineering application. Hence, they shall not be included in any fuel evaporation engineering calculations.

Chapter 10

Non-Luminous Flame Radiation Calculation

10.1. Importance of Non-Luminous Flame Radiation Calculation

For high pressure low emissions combustor, no matter which type of low emissions combustor design, with well-organized combustion there is always non-luminous flame. With the liner cooling design proposed in this book, cooling effectiveness is always unity, while convective heat transfer is from the hot wall to the cooling air, and not from the gas to the wall. That means non-luminous flame radiation is the only heat source to determine liner wall temperatures. If the combustor designer can calculate non-luminous flame radiation correctly, the liner wall temperature can be accurately predicted. That knowledge is very helpful for combustor design and development.

For a high FAR combustor, correct calculation of non-luminous flame radiation is also important. With correct non-luminous flame radiation, from the measured liner wall temperatures, the combustor developer will be able to correlate luminous factors as shown in Equ. 13 (this equation is only an example, because there needs more correlation). This correlation, nonetheless, will form a good foundation for high FAR combustor radiation calculations.

10.2. Non-Luminous Flame Radiation Calculation Method Shall Be Greatly Updated

This paragraph is copied verbatim from Lefebvre (1998). Lefebvre wrote the following: "Consider the radiation heat exchange between a gas at temperature T_g and the surface of a blackbody container at temperature T_w. While the black surface emits and absorbs heat at all wavelengths, the gas emits only a few narrow bands of wavelengths and absorbs only those wavelengths included in its emission bands." The net radiant heat transfer is given by:

$$R_1 = \sigma(\varepsilon_g T_g^4 - \alpha T_{w1}^4) \tag{19}$$

where

R_1 radiation heat transfer between combustion gas and wall

σ Stefan-Boltzmann constant $5.67 * 10^{-8}\ W/(m^2K^4)$

ε_g gas emissivity at T_g

α gas absorptivity at T_{w1}

T_g combustion gas temperature

T_{w1} liner wall inner surface temperature

Here ε_g relates to the emission of radiation from gas to wall and depends on T_g, but α_g applies to the absorption of radiation by gas from wall and hence depends on T_{w1}. In practice the surface exposed to the flame is not black, which has an effective absorptivity that is less than one. Lefebvre suggested using a factor $0.5(1+\varepsilon_w)$ to derive the following equation:

$$R_1 = 0.5 * \sigma * (1+\varepsilon_w) * (\varepsilon_g T_g^4 - \alpha_g T_{w1}^4) \tag{20}$$

Note Equ. 20 is not correct.

a Factor $0.5(1+\varepsilon_w)$ is not correct.
$0.5(1+\varepsilon_w)$ factor originates from Hottel's suggestion (Hottel and Egbert 1942). Originally, Hottel asserted that, ε_w is 0.8, thus using $0.5(1+\varepsilon_w)$ may cause an error to its maximum which will not be more than 10%. That means Hottel did not perform a serious research study. He merely

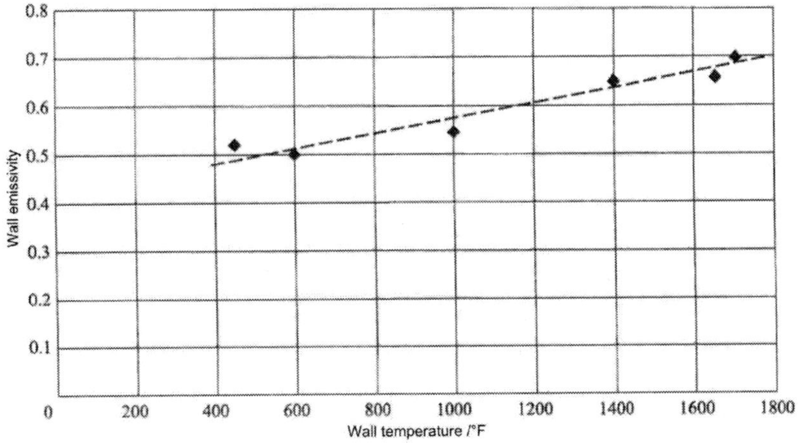

Figure 40. SS316, SS347 liner wall emissivity changes with wall temperature.

simply suggested this factor. In our more modern times we realize that liner wall emissivity is not always 0.8. In Lefebvre (1998) the author suggested that for stainless steel the ε_w is 0.8, and for mild steel it is 0.9. But this is not true. For SS316 and SS347 stainless steel the wall emissivity changes with wall temperatures as shown in Fig. 40. Even at a wall temperature of 1800°F, the emissivity is only 0.7. For Hastelloy X, Haynes 188, and Haynes 230, their wall emissivities are shown in Fig. 41. Fig. 41 may be expressed as:

$$\varepsilon_{w,\ \mathrm{Haynes}} = 0.5 + 5 * 10^{-4}(T_w - 600) \tag{21}$$

It is obvious that only with a temperature of 1800 F the wall emissivity is 0.8. If the wall is coated with $Y_2O_3 - ZrO_2$, its wall emissivity is shown in Fig. 42 (Liebert 1978). Coating emissivity depends upon coating thickness and coating surface temperature. Note the coating surface emissivity is always much less than 0.8.

b Even using a correct ε_w value, $0.5(1+\varepsilon_w)$ is still not correct. Edwards and Matavosian (1984) proved that although this factor is truly between ε_w and one, it is not at the middle. It is biased toward the ε_w side. Edwards and Matavosian (1984) showed that the value is $\varepsilon_w/(1-\Delta)$ and suggested

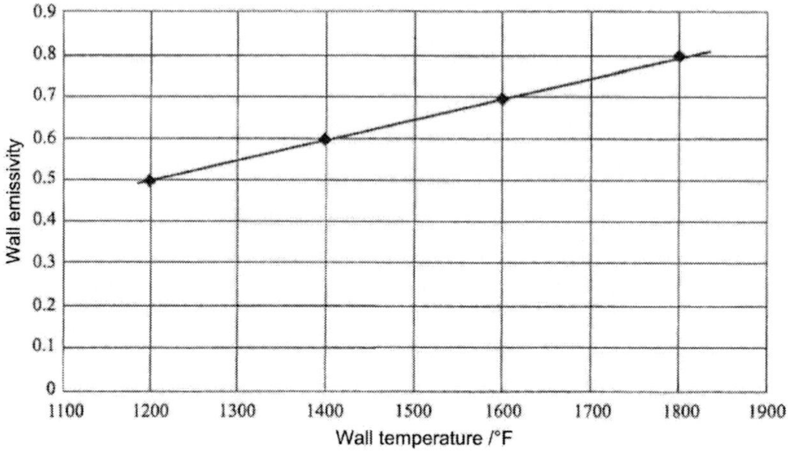

Figure 41. Hastelloy, Haynes 188, and Haynes 230 wall emissivities change with wall temperature.

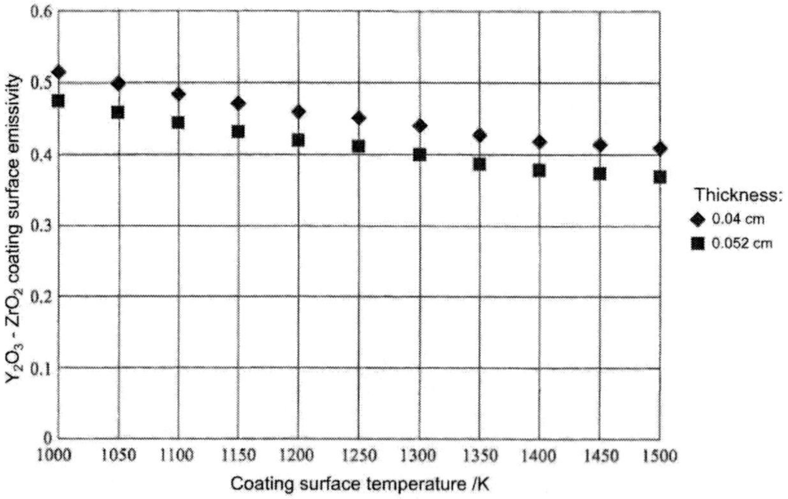

Figure 42. $Y_2O_3 - ZrO_2$ coating surface emissivity changes with coating surface temperature.

using a rather complicated method to calculate this factor. The present authors offer some simplification on this method to have the following factor:

$$\text{Factor} = \frac{\varepsilon_w}{1 - 0.17(1 - \varepsilon_w)} \tag{22}$$

The present authors have compared many situations using Equ. 22 and the complicated Edwards and Matavosian (1984) method, the maximum difference is 5%.

10.3. One Bar Combustion Pressure Water Vapor and Carbon Dioxide Gas Emissivity

Combustion gases in a gas turbine combustor consist mainly H_2O, CO_2, and N_2, while other gases such as CO, NO, NO_2, H_2, and O_2 comprise a very minor portion. N_2 is a major portion, but it emits no appreciable radiation even at very high temperatures. Thus, for non-luminous flame the most important consideration is radiation from H_2O and CO_2.

Gaseous radiation is quite different from solid radiation. Gaseous radiation is discrete, not a continuous band. For H_2O, the strongest emission bands are 1.9, 2.8, 6.7, and 21 μm. One strong band occurs at 2.7 μm. CO_2 has very strong bands at 4.3 and 15 μm. Also there are two strong emission bands near 2.7 μm. This means H_2O and CO_2 emission bands may overlap.

The values of H_2O and CO_2 emissivity can be obtained from the charts in Hottel and Egbert (1942) as shown in Fig. 43 (for H_2O) and Fig. 44 (for CO_2). In these two figures the vertical axis is emissivity, while the horizontal axis is temperature. Curves are for different products of partial pressure times beam length. Most importantly to note is that these curves are all for combustion pressure of only one atmospheric pressure. Later on in this book such emissivity will be named as $\varepsilon_{g,1\ bar}$. Combustion pressure is always higher than 1 bar, so the pressure effect must be included. Partial pressure is in *kpa*. From Edwards and Matavosian (1984) the partial pressure over combustion pressure for both water vapor and carbon dioxide can be expressed as:

$$\frac{\text{partial pressure}}{\text{combustion pressure}} = 2.1 * \frac{FAR}{1 + 1.05 * FAR} \tag{23}$$

Here it is assumed that water vapor has the same partial pressure as carbon dioxide, which is basically correct.

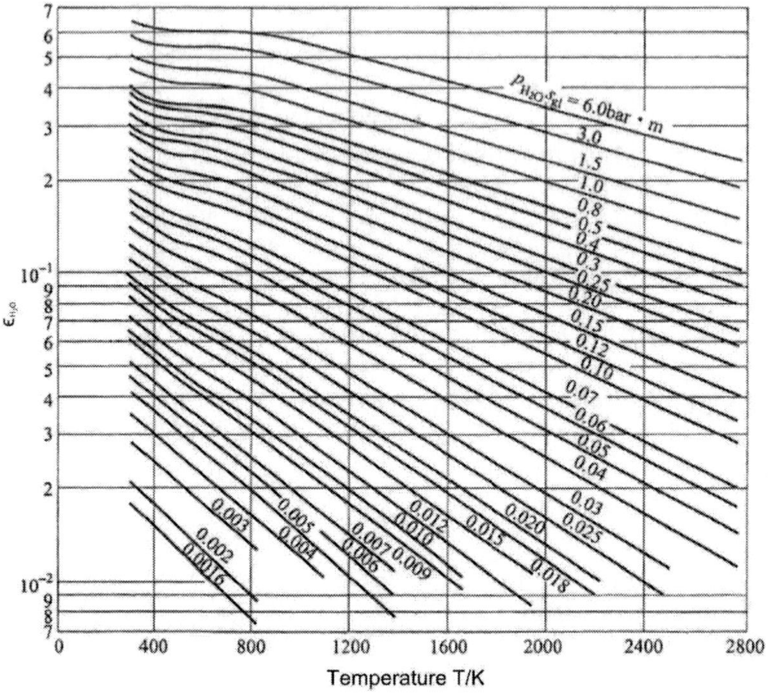

Figure 43. ε_{H_2O} versus temperature for values of $p_{H_2O}s_{gl}$, Hottel and Egbert (1942).

Beam length units are indicated in meter. From Mewes (1998), beam length may be calculated by:

$$\text{beam length} = 3.4 * \frac{\text{volume}}{\text{surface area}} \tag{24}$$

For any combustor the beam length should be calculated individually.

Fig. 43 and Fig. 44 are the foundation for non-luminous flame radiation calculations. The problem with Fig. 43 and Fig. 44 is that for engineering calculation, it is not convenient to read flame calculations directly from the figures. There need to be equations to calculate them. For water vapor emissivity the

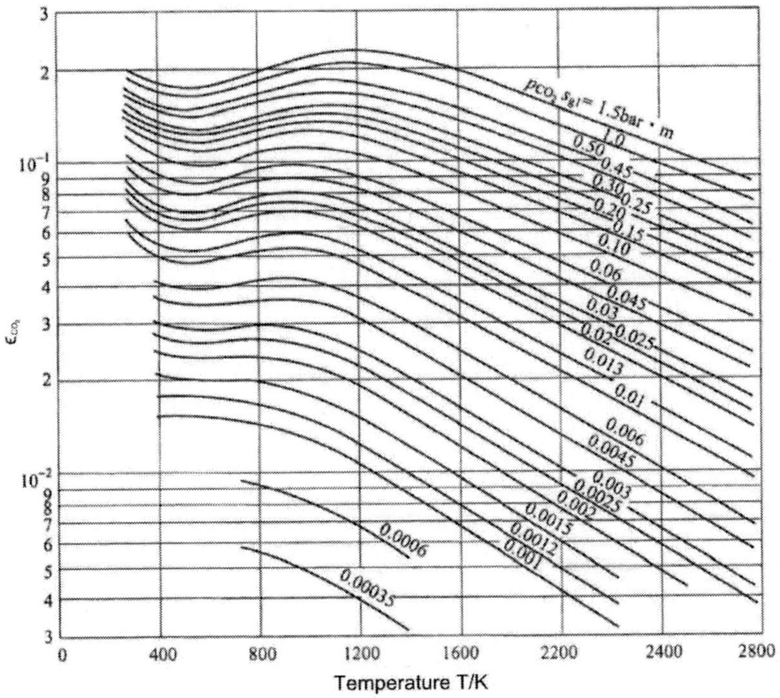

Figure 44. ε_{CO_2} versus temperature for various values of $p_{CO_2}s_{gl}$, Hottel and Egbert (1942).

following equation set may be used for calculations, (Mewes 1998).

$$\begin{aligned} \varepsilon_{H_2O} &= \exp(g_1 + g_2 u + h \tanh 3u) \\ g_{1,2} &= d + \frac{c_1 \ln(p_{H_2O}\, s_{gl}/z) + c_2\, p_{H_2O}\, s_{gl}}{1 + q(p_{H_2O} s_{gl})^2} \\ h &= \frac{0.119(p_{H_2O}\, s_{gl})^{0.568}}{0.342 + 0.993(p_{H_2O}\, s_{gl})^{0.568}} \\ u &= (T - 625)/625 \end{aligned} \tag{25}$$

The constants in above equations are shown in the following table

When calculate g_1, using column A.

When calculate g_2, using column B.

Table 5. Constants required for equation 25

	A	B
Constants	$g_1\ (p_{H_2O} s_{gl})$	$g_2\ (p_{H_2O} s_{gl})$
d	-0.370	-0.270
c_1	0.683	0.143
$c_2/(m \cdot bar)^{-1}$	-1.833	-0.233
$q/(m \cdot bar)^{-2}$	1.782	0.702
$z/(m \cdot bar)$	0.615	0.620

Carbon dioxide gas emissivity may be calculated by the following equation set (Mewes 1998).

$$\begin{gathered} \varepsilon_{CO_2} = a_0 + a_1\gamma + a_2\gamma^2 + a_3\gamma^3 \\ \gamma = \frac{(1273 - T_g)}{1000} \\ a_i = \frac{b_i(p_{CO_2} s_{gl})^{n_i}}{c_i + (p_{CO_2} s_{gl})^{n_i}} + (a_{\infty i} - b_i)\frac{(p_{CO_2} s_{gl})^{m_i}}{d_i + (p_{CO_2} s_{gl})^{m_i}} \end{gathered} \tag{26}$$

The constants in Equ. 26 are listed in the following table:

Table 6. Constants required for equation 26

i	a_∞	b_i	c_i	d_i	m_i	n_i
0	0.252	0.1166	0.04	0.477	1.542	0.802
1	0.01	0.0658	0.0245	1.712	0.25	0.715
2	-0.0955	-0.0535	0.013	0.115	2.45	1.076
3	-0.0303	-0.0806	0.0816	0.691	0.13	0.495

For utilizing Equ. 26 to calculate $\varepsilon_{CO_2,1\,bar}$, the use of constants must be explained:

1 If calculating a_0, the constants used are:
$a_\infty = 0.252$, $b_0 = 0.1166$, $c_0 = 0.04$, $d_0 = 0.477$, $m_0 = 1.542$, $n_0 = 0.802$

2 If calculating a_1, the constants used are:
$a_{\infty 1} = 0.01, b_1 = 0.0658, c_1 = 0.0245, d_1 = 1.712, m_1 = 0.25, n_1 = 0.715$

Then following the same method to calculate a_2, a_3.

Note that the present authors retain Fig. 43 and Fig 44 in this book and also present Equ. 25 and Equ. 26. Because when Equ. 25 and Equ. 26 are utilized to perform the calculation, errors may occur. Thus, the results can be verified with Fig. 43 and Fig. 44.

10.4. Pressure Effect on Water Vapor Emissivity

Since Fig. 43 and Fig 44 are for a combustion pressure of one bar, the combustion pressure effects on emissivity comprise two aspects:

a When combustion pressure is higher, water vapor and carbon dioxide partial pressure will be higher.

b The absorption bands broaden with pressure, the emissivity will become higher with a higher combustion pressure. This is the effect of pressure on emissivity.

Mewes (1998) reported the equations for calculation of pressure effects on carbon dioxide emissivity. The present authors calculated such effect, and our results show that the effect is very weak.

Mewes (1998) reported the equation for pressure effect on water vapor emissivity as follows:

$$f_{PH_2O} = 1 + (A-1)\exp\left\{-0.5\left[\log\frac{0.132(\frac{T_g}{1000})^2}{p_{H_2O}\,s_{gl}}\right]^2\right\}$$

$$A = \frac{(1.888 - 2.053\log\tau)p\left(1 + 4.9\frac{p_{H_2O}}{p}\sqrt{\frac{273}{T_g}}\right) + 1.10\left(\frac{T_g}{1000}\right)^{-1.4}}{0.888 - 2.053\log\tau + p\left(1 + 4.9\frac{p_{H_2O}}{p}\sqrt{\frac{273}{T_g}}\right) + 1.10\left(\frac{T_g}{1000}\right)^{-1.4}} \tag{27}$$

$$\tau = \frac{T_g}{1000},\ T_g > 750\ K,\ p < 100\ bar$$

where

p combustion pressure, *bar*

p_{H_2O} water vapor partial pressure, *bar*

The present authors use Equ. 27 to check the effect of T_g, s_{gl}, water vapor partial pressure, and combustion pressure on water vapor pressure correction. Results have shown that:

a Effect of beam length is weak.

b Water vapor partial pressure effect is weak.

c Most importantly is the temperature effect, as shown in Fig. 45. When temperature T_g is higher, the effect is weaker.

d The effect of combustion pressure is shown in Fig. 45 from 2 *bar* up to 90 *bar*. It is obvious that when combustion pressure is progressively rising, the curve is flattening. This is important because previously the pressure effect on gas emissivity was often overly estimated.

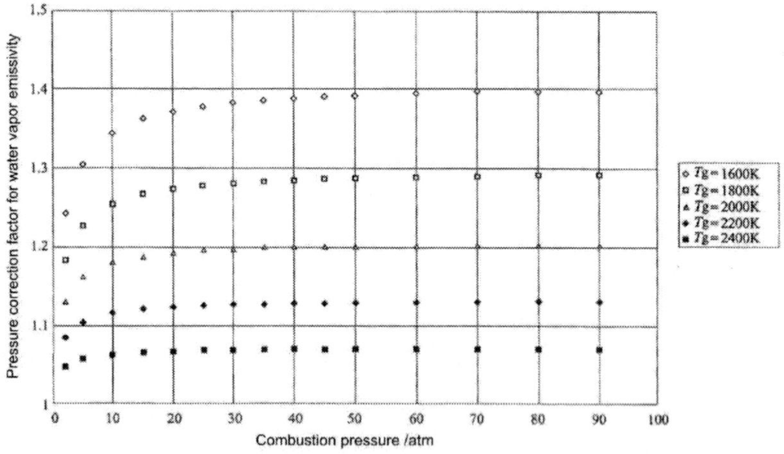

Figure 45. Pressure correction factors for water vapor emissivity at different combustion pressures and temperatures.

For next generation aero combustors, the pressure correction factor on water vapor emissivity is often between 1.15 to 1.25.

10.5. Gas Emissivity Overlapping

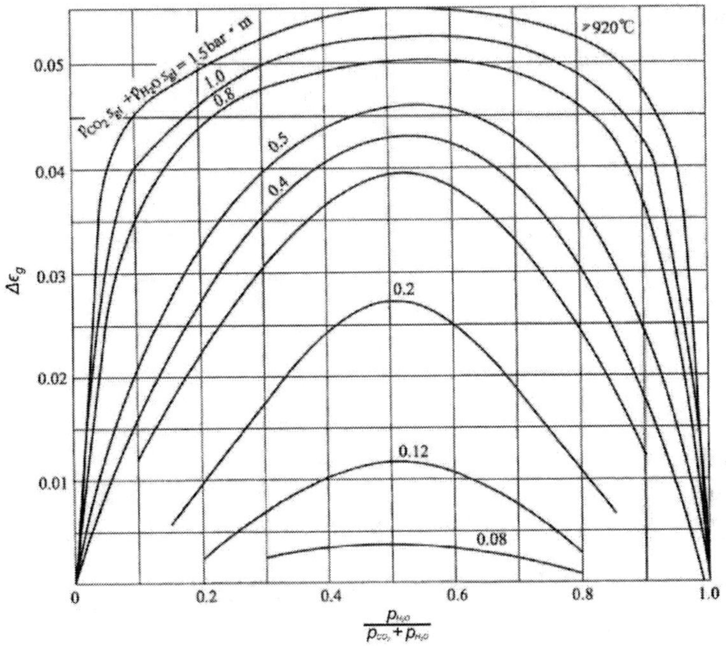

Figure 46. Overlapping correction ε_g for temperature $> 920\ ^\circ C$.

As previously mentioned, near the emission band 2.7 μm is an overlapping band of water vapor and carbon dioxide emissions. Thus, if both gases emit at the same time their emissivity will be less than the summation of the two gases emissivities when they emit individually. That is a correction of the overlapping effect, $\Delta\varepsilon_g$. The overlapping correction factor from Mewes (1998) is shown in Fig. 46. Note that this figure is suitable for temperature of 920°C and above. The water vapor partial pressure is very often 50% of the summation of water vapor and carbon dioxide partial pressure (except for JP-10), and so the correction $\Delta\varepsilon_g$ is always at the curve's top position. The present authors collected the median point of these curves and correlate them in an equation as follows:

$$\Delta\varepsilon_g = 0.00623 + 0.13233 * X - 0.1251 * X^2 + 0.03904 * X^3 \qquad (28)$$

where $X = (p_{H_2O} + p_{CO_2}) * s_{gl}$, $bar \cdot m$.

The comparison between Fig. 46 and Equ. 28 is shown in Fig. 47, and it is evident that the agreement is good. But for next generation aero combustors, and for high FAR combustor, the combustion zone fuel.air ratio is 0.068, whereas for a low emission combustor, although the combustion zone FAR is lower but combustion pressure can be 70 and above, then the summation of water vapor partial pressure and carbon dioxide partial pressure times beam length (for large liner) can be more than 1.5 $bar \cdot m$. $\Delta\varepsilon_g$ can be calculated by Equ. 28.

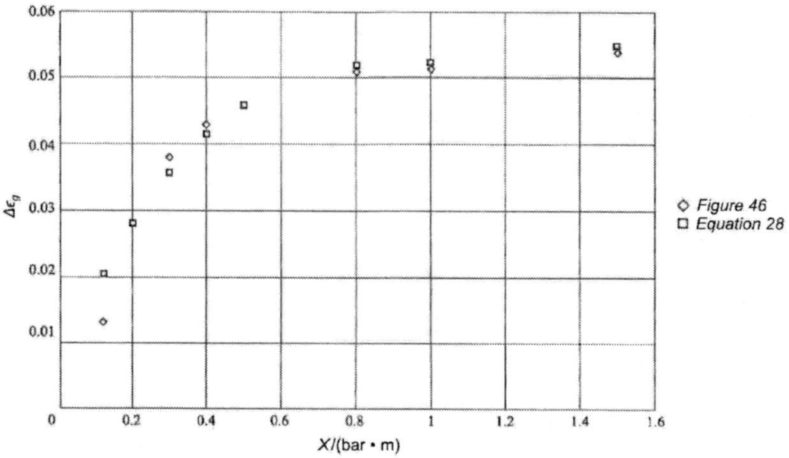

Figure 47. Comparison between correlation and calculation curve, $X = (p_{H_2O} + p_{CO_2}) \cdot s_{gl}$.

So, the non-luminous flame (combustion gas) emissivity then can be calculated by the following equation:

$$\varepsilon_g = \varepsilon_{H_2O,1\,bar} * f_{p,H_2O} + \varepsilon_{CO_2,1\,bar} - \Delta\varepsilon_g \tag{29}$$

where

$\varepsilon_{H_2O,1\,bar}$ calculated by Equ. 25

f_{p,H_2O} calculated by Equ. 27

$\varepsilon_{CO_2,1\,bar}$ calculated by Equ. 26

$\Delta\varepsilon_g$ calculated by Equ. 28

10.6. Absorptivity

In Lefebvre (1998), Lefebvre based on his research of decades past (Lefebvre and Herbert 1960), suggested utilizing the following equation:

$$\alpha_g/\varepsilon_g = (T_g/T_{w1})^{1.5} \tag{30}$$

Then Equ. 20 becomes:

$$R_1 = 0.5\sigma * (1+\varepsilon_w) * \varepsilon_g T_g^{1.5} * (T_g^{2.5} - T_{w1}^{2.5}) \tag{31}$$

Since Equ. 30 is incorrect, then Equ. 31 also is incorrect.

Let us consider one example. In Lefebvre (1998), page 279, he calculated $T_g = 2280\ K$, and with pressure, FAR, beam length, and luminous factor (1.7), substitute into the following equation:

$$\varepsilon_g = 1 - \exp\left[-290 * p * 1.7 * (FAR * s_{gl})^{0.5} * T_g^{-1.5}\right] \tag{32}$$

(Equ. 32 is in Lefebvre (1998), page 273).

to derive $\varepsilon_g = 0.61$. Note originally Equ. 32 was only for pressure lower than 5 *bar*. But from Lefebvre (1998), page 279 the example was for calculating a pressure of 30 *bar*, and then substituting ε_g into Equ. 31, to obtain wall temperature T_{w1} of 1283 K (page 297 bottom line). Equ. 31 has already included Equ. 30. Then using Equ. 30, one can insert the calculated T_g, T_{w1}, and ε_g values, and the result is $\alpha_g = 1.445$. This is obviously unreasonable, as absorptivity can never be over one. It easily can be seen that the power index in Equ. 30, which is 1.5, is wrong. Index 1.5 is only for a special case. From Edwards and Matavosian (1984), the following equation is recommended:

$$\alpha_g/\varepsilon_g = (T_g/T_{w1})^{0.5-b-a} \tag{33}$$

The following section explains how Equ. 33 is derived.

In classical radiation literature, there is always a statement regarding the emissivity of any certain gas and the absorptivity of that same gas. In the ε_g curve to read α_g, then using T_{w1} to replace T_g, and using $pa * s_{gl} * (T_{w1}/T_g)$ to replace pas_{gl}. Then multiply that with $(T_g/T_{w1})^{0.5}$. If expressed in the equation, it is:

$$\alpha_g(T_g,\ T_{w1},\ pas_{gl}) = (T_g/T_{w1})^{0.5} * \varepsilon_g(T_{w1},\ pas_{gl} * T_{w1}/T_g) \tag{34}$$

Here the bracket means a function of. Then we have:

$$\frac{\partial \ln \varepsilon_g}{\partial \ln (pas_{gl})} = a \tag{35}$$

$$\frac{\partial \ln \varepsilon_g}{\partial \ln T_g} = b \tag{36}$$

From Equ. 35 and Equ. 36 the following equation is true:

$$\varepsilon_g = constant * T_g^b * (pas_{gl})^a \tag{37}$$

Substitute Equ. 37 into Equ. 34, and we obtain:

$$\alpha_g = \left(\frac{T_g}{T_{w1}}\right)^{0.5} * constant * T_{w1}^b * \left(\frac{pas_{gl} * T_{w1}}{T_g}\right)^a \tag{38}$$

Equ. 38 divided by Equ. 37, we obtain:

$$\begin{aligned}\frac{\alpha_g}{\varepsilon_g} &= \left(\frac{T_g}{T_{w1}}\right)^{0.5} * \left(\frac{T_g}{T_{w1}}\right)^{-b} * \left(\frac{T_g}{T_{w1}}\right)^{-a} \\ &= \left(\frac{T_g}{T_{w1}}\right)^{0.5-b-a}\end{aligned} \tag{39}$$

Edwards and Matavosian (1984) calculated the values of b and a as shown in Fig. 48. Note that only when $k = 0.625$, $(0.5 - b - a) = 1.5$, this Equ. 30 is only a special case and generally cannot be used.

The combustor designer will experience difficulties to use Fig. 48 to determine $(0.5 - b - a)$. There needs to have an engineering method to determine $(0.5 - b - a)$. The present authors utilized the following method:

a After T_g has been determined, assume one value of $(T_g + 100)$ to calculate ε_{H_2O} and ε_{CO_2}, then assume another value of $(T_g{-}100)$ to calculate another set of ε_{H_2O} and ε_{CO_2}

b From ε_{H_2O} at $(T_g + 100)$ and T_g calculate one $\partial \ln \varepsilon_{H_2O}/\partial \ln T_g$, from T_g and $(T_g - 100)$ calculate yet another $\partial \ln \varepsilon_{H_2O}/\partial \ln T_g$, taking an average of these two its result is b_{H_2O}

c Calculate b_{CO_2} in the same way

d After calculating $p_{H_2O}s_{gl}$ and $p_{CO_2}s_{gl}$, assume one value of $(FAR + 0.005)$ and one value of $(FAR - 0.005)$, then calculate $p_{H_2O}s_{gl}$, ε_{H_2O}, $p_{CO_2}s_{gl}$, and ε_{CO_2}

e Calculate two $\partial \ln \varepsilon_{H_2O} / \partial \ln p_{H_2O}s_{gl}$ (one for $FAR + 0.005$ to FAR, and the other one is from FAR to $FAR - 0.005$), then take the average, this is a_{H_2O}

f Calculate a_{CO_2} in the same way

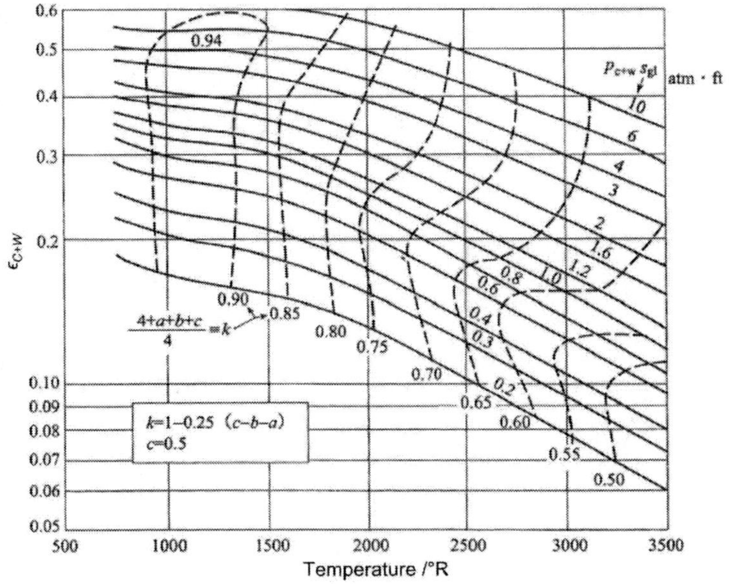

Figure 48. Calculated $(0.5 - b - a)$ from Edwards and Matavosian (1984). The units are rather different ε_{C+W} is emissivity summation of water vapor and carbon dioxide, p_{C+W} is summation of partial pressures, and beam length is in ft.

Now the issue is how to get a from a_{H_2O}, a_{CO_2}, and to get b from b_{H_2O}, b_{CO_2}.

Eckert (1969) suggested the following equation to calculate weighted a and

b for a mixture of water vapor and carbon dioxide:

$$a_g = \frac{a_{CO_2} * \varepsilon_{CO_2} + a_{H_2O,1\,bar} * f p_{H_2O}}{\varepsilon_{CO_2} + \varepsilon_{H_2O,1\,bar} * f p_{H_2O}} \tag{40}$$

$$b_g = \frac{b_{CO_2} * \varepsilon_{CO_2} + b_{H_2O,1\,bar} * f p_{H_2O}}{\varepsilon_{CO_2} + \varepsilon_{H_2O,1\,bar} * f p_{H_2O}} \tag{41}$$

Finally, the non-luminous flame radiation between hot gas and liner wall is calculated by the following equation:

$$R_1 = \frac{5.67 * 10^{-8} * \varepsilon_w * \left(\varepsilon_g T_g^4 - -\alpha_g T_{w1}^4\right)}{1 - 0.17 * (1 - \varepsilon_w)} \tag{42}$$

As a summary, in Lefebvre (1998), the calculation equations should be updated are as follows:

- $0.5 * (1 + \varepsilon_w)$ is not correct. It shall be $\varepsilon_w / [1 - 0.17(1 - \varepsilon_w)]$
- Pressure effect on CO_2 emissivity is weak
- At high combustion pressures, the pressure effect upon water vapor emissivity flattens out
- $\alpha_g / \varepsilon_g = (T_g / T_{w1})^{1.5}$ is incorrect; it should be $\alpha_g / \varepsilon_g = (T_g / T_{w1})^{0.5 - b - a}$
- All calculations should be performed by such equations.

References

Adkins, R. C., D. S. Matharu, and J. O. Yost. 1981. "The Hybrid Diffuser" [in en]. Publisher: American Society of Mechanical Engineers Digital Collection, *Journal of Engineering for Power* 103, no. 1 (January): 229–236. ISSN: 0022-0825, accessed December 7, 2020. doi:10.1115/1.3230702.

Bahr, D. W. 1987. "Technology for the Design of High Temperature Rise Combustors." Publisher: American Institute of Aeronautics and Astronautics _eprint: *Journal of Propulsion and Power* 3 (2): 179–186. Accessed December 7, 2020. doi:10.2514/3.22971.

Chin, J. S. 1995."Advanced Droplet Evaporation Model for Turbine Fuels." In *33rd Aerospace Sciences Meeting and Exhibit.* American Institute of Aeronautics / Astronautics. Accessed December 7, 2020. doi:10.2514/6.1995-493.

———. 1994a. "Advanced Spray Evaporation Model for Turbine Fuels." In *30th Joint Propulsion Conference and Exhibit.* American Institute of Aeronautics / Astronautics. Accessed December 7, 2020. doi:10.2514/6.1994-3277.

———. 1994b. "An Engineering Calculation Method for Multi-Component Stagnant Droplet Evaporation With Finite Diffusivity" [in en]. American Society of Mechanical Engineers Digital Collection, February. Accessed December 7, 2020. doi:10.1115/94-GT-440.

Chin, J. S., and D. Dang. 2021. "Design Considerations for Extra High-Pressure Ratio (70) Civil Aero Engine Low-Emission Combustor." In *AIAA Propulsion and Energy 2021 Forum.* American Institute of Aeronautics / Astronautics.

Chin, J. S., R. Durrett, and A. H. Lefebvre. 1984. "The Interdependence of Spray Characteristics and Evaporation History of Fuel Sprays" [in en]. Publisher: American Society of Mechanical Engineers Digital Collection, *Journal of Engineering for Gas Turbines and Power* 106, no. 3 (July): 639–644. ISSN: 0742-4795, accessed December 7, 2020. doi:10.1115/1.3239618.

Chin, J. S., and A. H. Lefebvre. 1983a. "Steady-state Evaporation Characteristics of Hydrocarbon Fuel Drops." Publisher: American Institute of Aeronautics and Astronautics _eprint: https://doi.org/10.2514/3.8264, *AIAA Journal* 21 (10): 1437–1443. ISSN: 0001-1452, accessed December 7, 2020. doi:10.2514/3.8264.

———. 1983b. "The Role of the Heat-up Period in Fuel Drop Evaporation." In *21st Aerospace Sciences Meeting.* American Institute of Aeronautics / Astronautics. Accessed December 7, 2020. doi:10.2514/6.1983-68.

Chin, Jushan. 2019. "Suggestions on High Temperature Rise Combustor." In *AIAA Propulsion and Energy 2019 Forum.* American Institute of Aeronautics / Astronautics. Accessed December 7, 2020. doi:10.2514/6.2019-4327.

Chin, Jushan, and Jianqin Suo. 2018. "Design of Aero Engine Lean Direct Mixing Combustor." In *2018 Joint Propulsion Conference.* American Institute of Aeronautics / Astronautics. Accessed December 7, 2020. doi:10.2514/6.2018-4921.

Dodge, L. G., and C. A. Moses. 1984. *Mechanisms of Smoke Reduction in the High-Pressure Combustion of Emulsified Fuels. Volume 3. Experimental Measurements and Computer Modeling of Evaporating Emulsified and Neat Fuel Sprays.* [in en]. Technical report. Section: Technical Reports. SOUTHWEST RESEARCH INST SAN ANTONIO TX, May. Accessed December 7, 2020.

Eckert, E. R. G. 1969. "Radiative Transfer, H. C. Hottel and A. F. Sarofim, McGraw-Hill Book Company, New York, 1967. 52 pages" [in en]. *AIChE Journal* 15 (5): 794–796. ISSN: 1547-5905, accessed December 7, 2020. doi:https://doi.org/10.1002/aic.690150504.

Edwards, D. K., and R. Matavosian. 1984. "Scaling Rules for Total Absorptivity and Emissivity of Gases" [in en]. Publisher: American Society of Mechanical Engineers Digital Collection, *Journal of Heat Transfer* 106, no. 4 (November): 684–689. ISSN: 0022-1481, accessed December 7, 2020. doi:10.1115/1.3246739.

Gauthier, J. E. D., M. F. Bardon, and V. K. Rao. 1991. "Combustion Characteristics of Multicomponent Fuels Under Cold Starting Conditions in a Gas Turbine" [in en]. American Society of Mechanical Engineers Digital Collection, March. Accessed December 7, 2020. doi:10.1115/91-GT-109.

Godsave, G. A. E. 1953. "Studies of the Combustion of Drops in a Fuel Spray—the Burning of Single Drops of Fuel" [in en]. *Symposium (International) on Combustion,* Fourth Symposium (International) on Combustion, 4, no. 1 (January): 818–830. ISSN: 0082-0784, accessed December 7, 2020. doi:10.1016/S0082-0784(53)80107-4.

Guin, C. 1999. "Characterisation of Autoignition and Flashback in Premixed Injection Systems." In *RTO Meeting Proceedings.*

Hottel, HC, and RB Egbert. 1942. "Radiant Heat Transmission from Water Vapor." *Transactions of the American Institute of Chemical Engineers* 38 (3): 0531–0568.

Lefebvre, A. H., and M. V. Herbert. 1960. "Heat-Transfer Processes in Gas-Turbine Combustion Chambers" [in en]. Publisher: IMECHE, *Proceedings of the Institution of Mechanical Engineers* 174, no. 1 (June): 463–478. ISSN: 0020-3483, accessed December 7, 2020. doi:10.1243/PIME_PROC_1960_174_039_02.

Lefebvre, Arthur H. 1998. *GAS Turbine Combustion, Second Edition* [in en]. Google-Books-ID: JJKQ9tj6q_8C. CRC Press, September. ISBN: 978-1-56032-673-1.

Liebert, Curt H. 1978. "Emittance and absorptance of NASA ceramic thermal barrier coating system.[for turbine cooling]." *NASA Technical Paper.*

McLellan, Charles H, and Mark R Nichols. 1942. *An Investigation of Diffuser-Resistance Combinations in Duct Systems.* Technical report. NATIONAL AERONAUTICS and SPACE ADMIN LANGLEY RESEARCH CENTER HAMPTON VA.

Mewes, D. 1998. "VDI-Wärmeatlas, 8. Auflage. VDI-Gesellschaft Verfahrenstechnik (Hrsg.), Springer-Verlag, Heidelberg 1997, 1374 Seiten, 980 Abb., 440 Tab., geb., DM 780,– ISBN 3-540-62719-7" [in en]. *Chemie Ingenieur Technik* 70 (12): 1642–1642. ISSN: 1522-2640, accessed December 7, 2020. doi:https://doi.org/10.1002/cite.330701234.

NASA, Glenn Research Center. 2004. *Technology Readiness Levels.*

Van Erp, C. A., and M. H. Richman. 1999. "Technical Challenges Associated with the Development of Advanced Combustion Systems" [in en]. In *RTO Meeting Proceedings.* ISBN: 978-92-837-0009-8, accessed December 7, 2020.

Yang, Gui Xiang, and Ju Shan Chin. 1990. "Experimental Study of the Effect of High Back-Pressure on the Atomization of a Plain Jet Injector Under Coaxial Air Flow." *Aerosol Science and Technology* 12, no. 4 (January): 903–910. ISSN: 0278-6826, accessed December 7, 2020. doi:10.1080/02786829008959402.

Authors' Contact Information

Jushan Chin

Senior Combustor Engineer
Rolls-Royce (Retired)
Previously Visiting Professor
Mechanical Engineering
Purdue University, AIAA Associate Fellow
Email: jschin2016@gmail.com

Jin Dang

Mechanical Engineer
Fossil Energy Research Corp

Index

D

E

F

U

V

W